Analytical Laser
Spectroscopy

NATO ASI Series

Advanced Science Institutes Series

A series presenting the results of activities sponsored by the NATO Science Committee, which aims at the dissemination of advanced scientific and technological knowledge, with a view to strengthening links between scientific communities.

The series is published by an international board of publishers in conjunction with the NATO Scientific Affairs Division

A	**Life Sciences**	Plenum Publishing Corporation
B	**Physics**	New York and London
C	**Mathematical and Physical Sciences**	D. Reidel Publishing Company Dordrecht, Boston, and Lancaster
D	**Behavioral and Social Sciences**	Martinus Nijhoff Publishers
E	**Engineering and Materials Sciences**	The Hague, Boston, and Lancaster
F	**Computer and Systems Sciences**	Springer-Verlag
G	**Ecological Sciences**	Berlin, Heidelberg, New York, and Tokyo

Recent Volumes in this Series

Series B: Physics

Analytical Laser Spectroscopy

Edited by

S. Martellucci

The Second University
Rome, Italy

and

A. N. Chester

Hughes Aircraft Company
El Segundo, California

Plenum Press
New York and London
Published in cooperation with NATO Scientific Affairs Division

Proceedings of a NATO Advanced Study Institute on
Analytical Laser Spectroscopy,
which was the Ninth Annual Course of the
International School of Quantum Electronics,
held September 23–October 3, 1982,
at the Ettore Majorana Center for Scientific Culture,
Erice, Sicily, Italy

Library of Congress Cataloging in Publication Data

NATO Advanced Study Institute on Analytical Laser Spectroscopy (1982: Ettore
    Majorana International Centre for Scientific Culture)
    Analytical laser spectroscopy.

    (NATO ASI series. Series B, Physics; v. 119)
    Proceedings of a NATO Advanced Study Institute on Analytical Laser Spectro-
scopy, which was the 9th annual course of the International School of Quantum
Electronics, held in Erice, Sicily, 9/23–10/31/82, at the Ettore Majorana Center for
Scientific Culture.
    "Published in cooperation with NATO Scientific Affairs Division."
    Bibliography: p.
    Includes index.
    1. Laser spectroscopy—Congresses. I. Martellucci, S. II. Chester, A. N. III. Title.
IV. Series.
QC454.L3N38   1982   535.5′8   84-26371

ISBN-13: 978-1-4612-9484-9   e-ISBN-13: 978-1-4613-2441-6
DOI: 10.1007/978-1-4613-2441-6

©1985 Plenum Press, New York
Softcover reprint of the hardcover 1st edition 1985

A Division of Plenum Publishing Corporation
233 Spring Strèet, New York, N.Y. 10013

FOREWORD

This volume contains the Proceedings of a two-week NATO A.S.I. on "Analytical Laser Spectroscopy", held from September 23 to October 3, 1982 in Erice, Italy. This is the 9th annual course of International School of Quantum Electronics organized under the auspices of the "E. Majorana" Center for Scientific Culture.

The Advanced Study Institute has been devoted to the analytical applications of lasers in spectroscopy. Atomic and molecular spectroscopy is one of the research fields in which the use of lasers has had a dramatic impact. New spectral information, difficult or impossible to gather by classical spectroscopy, extremely high resolution spectroscopy of atoms and molecules made possible by the overcoming of the Doppler effect, selective excitation and detection of single atomic and molecular quantum states are just few typical examples of how laser sources have revolutionized the field, offering challenging problems of both fundamental and applied nature.

Among the possible approaches to a course on Analytical Laser Spectroscopy, the one which emphasizes the scientific and technological aspects of the advanced laser techniques when applied to chemical analysis has been chosen. In fact, it reflects the new policy of the School to stress the advanced scientific and technological achievements in the field of Quantum Electronics. Accordingly, the course has given the broadest information on the ultimate performances of analytical laser spectroscopy techniques and the perspectives of their applications.

Because of the great variety of applications in chemistry, biology, engineering and related branches of science, this school on Analytical Laser Spectroscopy addressed a subject of interdisciplinary interest. The formal sessions have been balanced between tutorial presentations and lectures focusing on unsolved problems and future directions. In addition, wide time has been provided for the participants to meet together informally for additional technical discussions on laser instrumentation.

This NATO Advanced Study Institute provided not only a thorough tutorial treatment of the field, but also, through panel discussions and additional lectures, treated topics at the forefront of current work. Therefore the character of the course was a blend of current research and tutorial reviews. Many of the world's acknowledged leaders in the field were brought together to review and speculate on the accomplishments of Analytical Laser Spectroscopy.

The severe time deadline, established in order to achieve the most convenient publication date, did not allow the inclusion of the contributions by some very busy authors (as for instance, A. Mooradian and B. Stoicheff).

The NATO A.S.I. has taken advantage from the very active audience; most of the students were experts in the field of Analytical Laser Spectroscopy and contributed with panel discussions and seminars. Some students' seminars are also included in the "contributed papers" section of these Proceedings.

This volume would, of course, not be possible without the considerable efforts put forth by the authors represented here. Among them, N. Omenetto and M. A. Biancifiori volunteered to edit the manuscripts in Erice and are cordially acknowledged. We also wish to mention with sincere thanks the very specialized assistance of Mrs. M. Fiorini in the successful organization of the course and for her competent and timely editorial work on the entire collection of papers. In addition we must acknowledge the very qualified collaboration offered by Mrs. M.T. Petruzzi and Mrs. M.F. Tani.

Finally, but not the least of all, we are also indebted to Dr. R.H. Andrews of Plenum Press for his assistance in the preparation of this volume.

The Directors of the I.S.Q.E.

A.N. Chester   S. Martellucci
Hughes A.C.   The Second Iniversity
EL Segundo, CA (U.S.A.)   Rome (Italy)

April 5, 1983

CONTENTS

CONTRIBUTED PAPERS

LASER DOUBLE-RESONANCE SPECTROSCOPY

IN MOLECULAR PHYSICS

W. Demtröder

Fachbereich Physik
Universität Kaiserslautern
Kaiserslautern, Fed. Rep. Germany

I. INTRODUCTION

Detailed experimental investigations of electronic transitions in diatomic or polyatomic molecules and closer examinations of excited molecular states are often impeded by fundamental difficulties: The visible and ultraviolet spectra of many molecules show an exceedingly high spectral line density where many lines overlap within their Doppler-width. The application of "classical", Doppler-limited spectroscopic techniques to such cases often yields quasicontinuous spectra concealing many finer details such as rotational structure or fine- and hyperfine splittings. Furthermore, the excited states of molecules often have a perturbed level structure which results in congested and irregular spectra and makes their assignment a tedious task.

On the other hand a more thorough knowledge of excited molecular states is highly desirable since they play an important role in chemical reactions. All light-induced reactions, as for example photobiological processes, proceed via electronically excited states. Although studies of laser induced chemical reactions have gained increasing interest, our knowledge of the detailed structure of excited states is still insufficient.

There are several ways to overcome the experimental difficulties. Either the spectral resolution can be increased beyond the limit of the Doppler-width, or the number of lines has to be reduced. The first solution is based on Doppler-free spectroscopic techniques such as saturation spectroscopy[1], polarization spectroscopy[2] or linear laser spectroscopy in collimated molecular beams[3]. The reduction of the number of thermally populated absorbing levels and

1

thus of the density of lines may be achieved by a drastic decrease of the temperature, down to a few degrees (°K). Molecular spectroscopy at such low temperatures can be realized for instance in supersonic molecular beams where adiabatic cooling during the expansion from a high pressure region into the vacuum results in rotational temperatures down to below 1°K[4].

Laser double resonance spectroscopy is capable of combining both advantages for the investigation of complex molecular spectra: They allow Doppler-free resolution and they reduce the number of detected molecular transitions to those few which start from a single selected level. Many different double resonance schemes have been applied so far to molecular spectroscopy, using either two lasers or one laser and a microwave or radiofrequency field.

Figure 1 illustrates schematically three possible schemes of optical-optical double resonance (OODR) where two independent lasers are utilized: a pump laser which is stabilized on a selected transition between two levels i and k and a tunable probe laser which is scanned through the spectral range of interest and monitors all those transitions connected either with level i or with level k. The beams from the two lasers may travel collinear or anti-parallel through the sample.

The pump laser "labels" both levels i and k either by changing their populations N_i, N_k due to optical pumping or by altering the orientation of molecules in these levels. Both effects will change the absorption or induced emission of the probe laser or its state of polarization which can be monitored with appropriate detection techniques.

Pulsed lasers as well as cw lasers can be used. Both types have their advantages and disadvantages as will be discussed below. In case of cw lasers it is convenient to chop the pump intensity at a frequency f with a period 1/f long compared to the relaxation time of the pumped level. This guarantees quasi-stationary conditions during the pump cycle. The populations N_i and N_k are then switched at the frequency f between the thermal populations $N_i(0)$, $N_k(0)$ with the pump laser off and the values $N_i(L)$, $N_k(L)$ with the pump laser on. When the probe laser absorption is monitored at the frequency f with phase sensitive detection techniques, only those transitions are detected which are connected either with level i or with level k.

We will now discuss the three schemes of Figure 1 in more detail and illustrate them by several examples.

II. OPTICAL-OPTICAL DOUBLE RESONANCE WITH A COMMON LOWER LEVEL

The double resonance scheme of Figure 1a may be regarded as an inversion of laser induced fluorescence (see Figure 2). The LIF

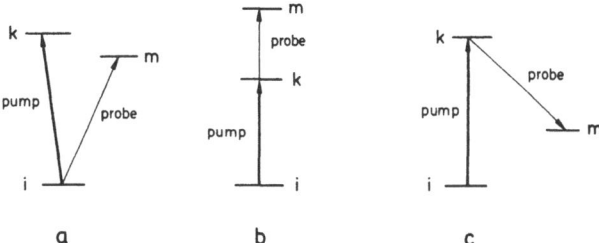

Fig. 1.   Three possible OODR-schemes.

method is based on the selective population of a vibrational rotational level (v'_k, J'_k) in an excited electronic state. The fluorescence spectrum, emitted from this level is simpled and readily
assigned. It gives information about the levels $m = (v'', J'')$ of the
lower state on which the fluorescence terminates. The double resonance signals, on the other hand, consist of all allowed transitions
from the lower level (v''_i, J''_i), labelled by the pump laser, to all
accessible levels (v'_m, J'_m) of upper electronic states. The upper
levels k and m of pump and probe transitions may belong to the same
electronic state or to different states.

The method is of particular advantage when the upper state
reached by the probe laser is perturbed. In this case many lines
will be shifted from their regular position and it is by no means
trivial to assign such lines in a normal absorption spectrum, even if
all lines are resolved. The double resonance signals, however, are
much easier to identify since the lower level (v''_i, J''_i) of each probe
transition is already known from the pump transition. The selection
rules for optically allowed transitions leave only a few choices for
upper levels (v'_m, J'_m).

In diatomic molecules, for example, the selection rule $\Delta J = \pm 1,0$
for the rotational quantum number J allows only rotational levels
with $J'_m = J''_i \pm 1$ or $J'_m - J''_i$ to be reached by the probe laser[5]. If
the lower electronic state i and the upper state m both are Σ-states,
only transitions with $\Delta J = \pm 1$ are allowed and the spacings of the two
double resonance signals to levels $(v', J''_i + 1)$ and $(v', J''_i - 1)$
immediately yield the rotational spacings in the upper state.

When pulsed lasers with larger bandwidth than that of single
mode cw lasers are applied to double resonance spectroscopy of complex spectra, several molecular transitions may fall within the
bandwidth of the pump laser, which therefore simultaneously labels
several levels $i_1, i_2 \ldots$. The probe laser then monitors all accessible transitions from these levels and the double resonance spectrum
may loose some of its simplicity. Even when single mode cw lasers
are used in gas or vapor cells with Doppler-broadened absorption
lines, all transitions which overlap with the pump laser wavelength

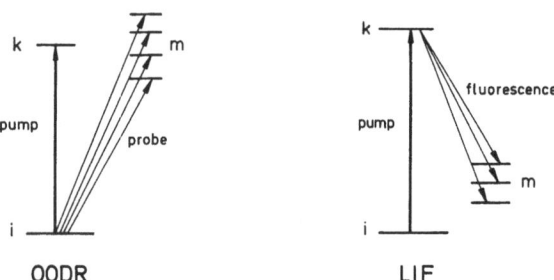

Fig. 2.   Comparison of OODR-spectroscopy (left) with laser induced
          fluorescence (LIF)-spectroscopy (right).

within the Doppler-width can be pumped and give rise to double resonance signals. The intensities of the double resonance signals depend
on the change $\Delta N = N_i(0) - N_i(L)$ induced by the pump laser. Those
transitions which overlap only on the far wings of their Doppler
profiles with the pump laserwavelength will be therefore less intense
than the selected transition which coincides exactly with the pump
laser.

There are two further effects which increase the number of
possible double-resonance signals. These are due to fluorescence
from the pumped upper level k and to collisional energy transfer
between the pumped lower level i and neighboring levels (v", J")
(Figure 3). The fluorescence from level (v_k', J_k') populates many
lower levels (v", J") (see Figure 2b), according to the Franck-Condon
factors of the fluorescence transitions $(v_k' \rightarrow v'')$. Because the
spontaneous lifetime of the upper level is generally much shorter
than the chopping period 1/f, the phase of the population modulation,
caused by the chopped pump laser, is for these levels (v", J") the
same as for level k, but opposite to that of level i.

Since the pump laser depletes the population N_i of level i below
the thermal equilibrium value $N_i(0)$, collision induced transitions
from neighboring levels to i will more frequently than collisions
which depopulate level i. This means that the populations N(v",J")
of these adjacent levels will also decrease during the period where
the pump laser is on. The population decrease ΔN_j of a level $(v_j'',
J_j'')$ depends on the population change $\Delta N_i = N_i(0) - N_i(S)$, on the
collision cross section σ_{ij} and the density of colliding molecules.
Measurements of the ratios $\Delta N_j / \Delta N_i$ can be used to determine these
cross sections (see below).

For high resolution spectroscopy of dense spectra all these
effects may impede the assignment of the double resonance signals.
They can be eliminated by combining the OODR-method with Doppler-free
techniques under collision free conditions. Two examples shall
illustrate this.

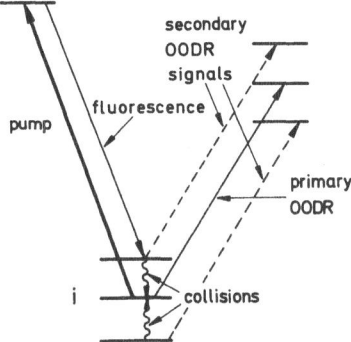

Fig. 3. Generation of "secondary" double resonance signals by pump-
         laser induced fluorescence and by collisional population
         transfer.

The first example demonstrates the application of OODR spectroscopy to the assignment of lines in the extremely complex visible
spectrum of NO_2. The experimental arrangement is shown in Figure 4.
The beams from two single mode lasers cross the collimated molecular
beams perpendicularly. Both beams pass through a chopper wheel with
two different rows of holes. The pump beam intensity I_1 is chopped
at a frequency f_1, the probe beam intensity I_2 at a frequency f_2.
The fluorescence induced by both lasers is used to monitor the absorption. The population density N_i of molecules, having passed
through the pump beam, is

$$N_i(L) = N_i^0(1 - a \cdot I_1) \qquad (1)$$

fluorescence intensity induced by the probe laser, which crosses the
molecular beam a few mm downstreams from the pump laser is given by

$$I_{f1} = b \cdot N_i(L) \cdot I_2 = b \cdot I_2 \cdot N_i^0 - a \cdot b \cdot N_i^0 \cdot I_1 I_2 \qquad (2)$$

$$= b\, N_i^0\, I_{20}(1-\cos 2\pi f_2 t) - ab N_i^0 I_{10} I_{20}(1-\cos 2\pi f_1 t)(1-\cos 2\pi f_2 t)$$

The first term represents the fluorescence induced by the probe laser
in the absence of the pump laser. The second term describes the
decrease in fluorescence due to the saturation of the population N_i
caused by the pump laser. Evaluation of the product yields:

$$I_{f1} = b(1-aI_{20})I_{10}(1-\cos 2\, f_2 t) + ab N_i^0 \cos 2\, f_1 t$$

$$- \frac{1}{2}\, ab N_i^0 I_{10} I_{20}\, [\cos 2\pi(f_1 + f_2)t + \cos 2\pi(f_1 - f_2)t]$$

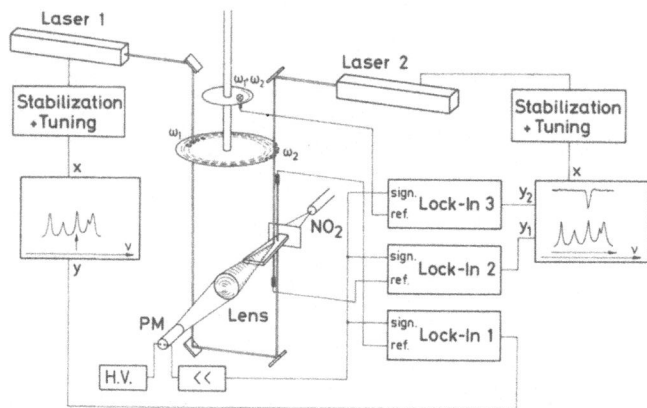

Fig. 4.   Experimental arrangement for sub-Doppler OODR-spectroscopy
          in a molecular beam under collision-free conditions. The
          two lasers are chopped at different frequencies f_1 and f_2
          and the OODR-signals are monitored through a lock-in ampli-
          fier, tuned to the sum frequency.

If the fluorescence is monitored through a lock-in amplifier, tuned
to the sum frequency $(f_1 + f_2)$ only the double resonance signals are
detected, while the fluorescence, induced by the pump laser or the
probe laser alone can be monitored at the frequencies f_1 and f_2
respectively.

     Figure 5, which shows a section of the excitation spectrum of
NO_2 around $\lambda = 488$ nm, illustrates the complexity of the spectrum[6].
There is no regular pattern to be recognized. The OODR-method is
helpful in assigning the spectrum. The lower trace of Figure 6
reproduces a small part of the spectrum of Figure 5. If the pump
laser is stabilized on line No. 4 and the probe laser is tuned[7],
the OODR-spectrum, shown in the upper trace is obtained at the sum
frequency, which proves that lines No. 1 and No. 4 share a common
lower level. The separation of these two OODR-signals represent the
energy separation of two upper levels m_1 and m_2, which are both
optically connected with the common ground state level. Further
studies showed that also the line pairs 2 and 5 and 3 and 6 were
coupled by a common lower level, and that the lines 1, 2, 3 and 4, 5,
6 represent the three hyperfine structure components of two rotational transitions $(J_i'', K_i'') \rightarrow (J_m', K_m')$ with m = 1, 2.

     Since the Doppler-width is reduced by a factor $\varepsilon \cong 100$, which
equals the collimation ratio of the molecular beam, simultaneous
excitation of several pump transitions can be eliminated. At the low
pressures in the molecular beam collisions can be neglected. Therefore the combination of Doppler-free excitation with double-resonance
techniques simplifies the OODR-spectrum and makes its assignment
straight forward, provided the pump transition is known.

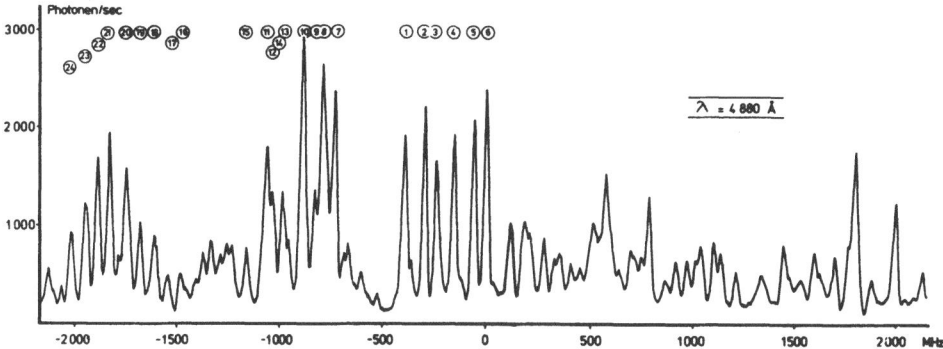

Fig. 5. Sub-Doppler excitation spectrum of NO_2 around λ = 488 nm.

III. STEPWISE EXCITATION

The second OODR-scheme of Figure 1b has proved to be very successful for the study of highly excited molecular states. The first
laser populates a selected level in an excited state and the second
laser induces transitions from this level into higher states. The
final levels m reached by absorption of two photons, have the same
parity as the initial ground state level i. This OODR excitation,
which may be regarded as a resonant two photon transition, therefore
gives complementary information compared with ultraviolet absorption
spectroscopy where states with comparable energy but opposite parity
may be reached by absorption of one ultraviolet photon.

The second transition K → m of the OODR can be monitored in
different ways: when the fluorescence induced by the second laser is
detected as a function of the wavelength λ_2 of the second laser, one
obtains the excitation spectrum of the transitions K → m. If the
states m are high lying molecular Rydberg-states, they may autoionize
and the number of ions detected as a function of λ_2 is a measure for
the transition K → m. Another method, called "polarization labelling" is based on the orientation of the optically pumped molecules
in level K. If the sample is placed between two crossed polarizers,
linearly polarized light will only pass the second polarizer if its
plane of polarization is being turned through the interaction with
these oriented molecules [see reference 2 on polarization spectro-
scopy]. Only those wavelengths are therefore transmitted to the
detector which coincide with transitions K → m or i → m. If the
probe light is broadband radiation, generated for example by a broadband nitrogen laser pumped dyelaser, all transitions k → m within the
bandwidth of the probe can be detected simultaneously with a photo-
plate or an optical multichannel analyzer.

The intermediate levels K have radiative lifetimes ranging from
a few ns to several μs. When pulsed lasers with pulse durations of

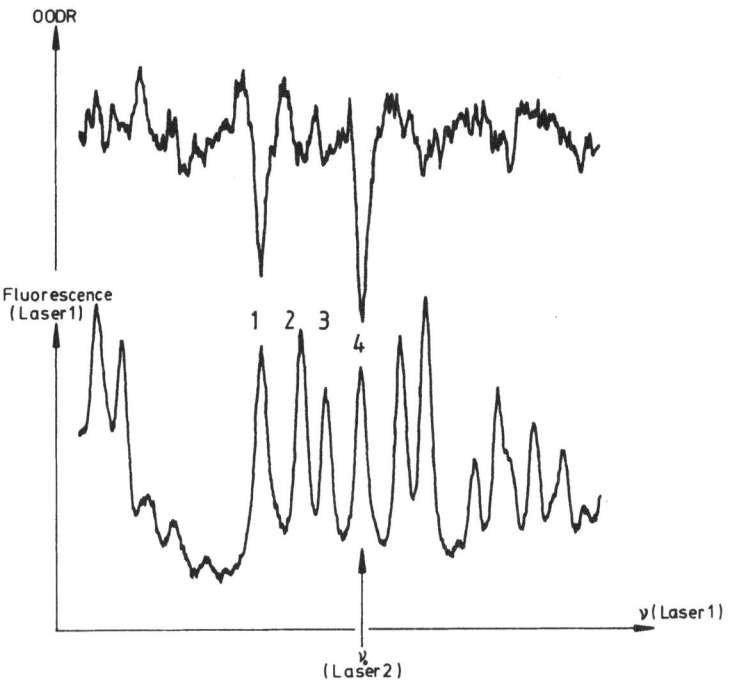

Fig. 6.   OODR-spectroscopy of NO_2. Lower spectrum: section of the
          sub-Doppler spectrum of Figure 5, obtained when laser L1 is
          tuned. Upper spectrum: OODR-signals obtained with the pump
          laser L1 stabilized on line No. 4, and L2 being tuned.

1 ÷ 10 nsec are used, typical for nitrogen laser pumped dyelaser,
even the shorter lifetimes of some nsec raise no problems. The
double resonance signal S is given by

$$S = a \cdot N_i \cdot B_{iK} \cdot I_1 \cdot B_{Km} \cdot I_2 \cdot \Delta V \tag{4}$$

as long as saturation is avoided. Since the molecules do not travel
far during the short pump pulse duration, the signal is proportional
to the density N_i times the pumped volume ΔV.

The peak powers of pulsed dye lasers readily reach several KW
and are therefore sufficiently high to saturate the transition i → K.
This implies that nearly 50% of the unsaturated initial population N_i
can be transferred to level k. The probability P_{km} for further
excitation to higher levels m by the second laser with energy density
ρ_2 at the location of the sample is

$$P_{km} = \frac{B_{km} \, \rho_2}{B_{km} \, \rho_2 + \Sigma \, A_{kn}} \tag{5}$$

where B_{km} is the Einstein coefficient for absorption on the transition $k \to m$ and ΣA_{kn} represents the transition probability for spontaneous decay of level k into lower levels n.

Although the transition probability for transitions $k \to m$ to higher Rydberg levels m decreases with increasing principal quantum number of these levels, pulsed lasers have sufficiently high energy densities ρ to make $B_{km} \rho_2 >> \Sigma A_{km}$. The probability P_{km} then approaches unity. Under such favorable conditions the detection of OODR-signals imposes no serious problems.

The situation is different when cw lasers are used. Since their output power are much smaller than the peak powers of pulsed lasers, the probability P_{km} may be much smaller than unity, which means that only a small fraction of all molecules in level k can be further excited.

The application of cw lasers to stepwise excitation of molecules faces another problem, related to the short spontaneous lifetime τ of many molecular levels k. Let the two collinear laser beams have a common diameter d. When a molecule in level i diffuses into the pump beam, it may be excited into level k at a position x. During its spontaneous lifetime τ it travels a mean distance $\bar{s} = \tau\bar{v}$, where $\bar{v}$ is the mean thermal velocity. Only that small part of the spatially distributed probe laser power between x and $x + \bar{s}$, that is "seen" by the excited molecule before it decays, is useful for further excitation $k \to m$. After the molecule has spontaneously decayed, it is lost for the pumping processes because only the small fraction $\eta = A_{ki} / \underset{n}{\Sigma} A_{kn}$ of all spontaneous decays terminates on the initial level i and can be pumped again. Most of the molecules decay into other levels n which are not optically connected with level i and could only return to it by collisional transfer.

In order to illustrate this problem, let us consider a typical example: with $\bar{v} = 500$ m/sec and $\tau = 10^{-8}$ sec we obtain $\bar{s} = 0.5$ μm. When the two focussed laser beams have beam waists of $w = 0.1$ mm, the excited molecule travels on the average less than 0.5% of the probe beam diameter. This means that only a small fraction of the spatially distributed probe power is useful to induce the double resonance transition $k \to m$.

For OODR-spectroscopy in molecular beams with cw lasers the geometrical arrangement of the two laser beams is of crucial importance for the optimization of the OODR-signals. The intensity of the first laser should be sufficiently high to ensure that all molecules which pass through the laser beam, can be excited. On the other hand, however, it should not exceed an upper limit, to prevent that the molecules are already excited before the maximum of the Gaussian beam profile where the second laser has not sufficient power to compete with the spontaneous decay. In any case the second laser should have more power than the first and both diameters should be not much larger than the mean path $\bar{s}$. Since the molecular beam width is of the order of 1 mm, the focussing by spherical lasers would make the laser beam diameter much smaller than the molecular beam diameter and most of the molecules would pass outside the laser beam without being pumped. The most efficient geometrical arrangement for two step excitation with cw lasers in molecular beams uses a combination of spherical and cylindrical lenses[8,9] to focus the two collinear laser beams into a thin sheet of light with an elliptical cross section at the intersection with the molecular beam (see Figure 7).

IV. TWO STEP RESONANCE IONIZATION SPECTROSCOPY

The optimization of stepwise excitation is of particular relevance for the development of a very sensitive detection method, called two photon resonance ionization spectroscopy (see the paper by G.S. Hurst in this volume). This method is based on the OODR-scheme of Figure 1b where the upper states m are either continuous states above the ionization limit or high lying molecular Rydberg-states which can autoionize. The molecular ions generated by the second laser on the transition $k \rightarrow m = k \rightarrow I$ are used to monitor the absorption of the first laser since they are only produced from excited states k, populated by the first laser.

In this scheme the wavelength of the second laser is kept fixed and that of the first laser is tuned. Any time the first laser wavelength coincides with a transition $i \rightarrow k$, the second laser produces ions, provided $h\nu_2 > (IP - E_k)$. The number of ions n_{ion} produced per second, is

$$n_{ion} = N_k \cdot P_{KI} = n_a \frac{P_{kI}}{P_{kI} + \sum_n A_{kn}} =$$

$$= n_a \frac{B_{kI}\, \rho_2}{B_{kI}\, \rho_2 + \Sigma A_{kn}} = \qquad (6)$$

$$= N_i \cdot \sigma_{ik} \cdot (P_1/h\nu_1) \cdot \Delta x\, \frac{B_{kI}\, \rho_2}{B_{kI}\, \rho_2 + A_k}$$

where n_a is the number of pump laser photons $h\nu_1$, absorbed per second at a pathlength Δx on the transition $i \rightarrow k$ at an incident power P_1, and B_{kI} is the Einstein coefficient for the transition from level k into the ionization state and $A_K = \Sigma A_{kn}$ is the total relaxation rate of level k. Equation (6) shows, that the rate at which the ions are produced is proportional to the rate n_a of absorbed photons $h\nu_1$,

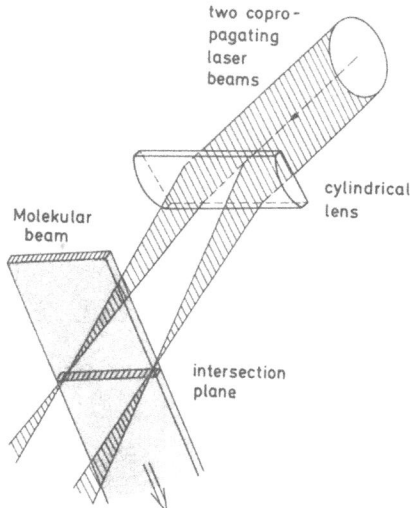

Fig. 7. Optimum focussing of laser beams for stepwise excitation
         with cw lasers in molecular beams.

which on the other hand is proportional to the absorption cross
section σ_{ik}.

     This resonance ionization spectroscopy is by far more sensitive
than excitation spectroscopy as the following short estimate shows:
Excitation spectroscopy is based on measuring the rate of fluorescence photons n (λ_1) emitted from the excited state k, which was
populated by a laser with wavelength λ_1. The fluorescence is emitted
into a solid angle 4π. Therefore, only a fraction ϵ of the fluorescence intensity can be collected onto the photomultiplier cathode.
The quantum yield η of the photocathode is about $\eta \leq 0.25$. The
number of absorbed photons necessary to produce one detectable photo-
electron is then, with $\epsilon = 0.05$ and $\eta = 0.2$:

     $n_a \geq 1/(\eta \cdot \epsilon) \cong 100$

Even cooled photomultipliers have a thermal dark current of at least
10 electrons/sec. In order to obtain a signal well above the dark
current, a minimum absorption rate of $n_a \geq 10^3$/sec is necessary.

     In case of ionization spectroscopy the collection efficiency for
the photo-ions can be as large as $\epsilon = 1$. According to Equation (6)
the number of absorbed pump-photons necessary to produce one ion is n_a
$= (P_{KI} + A_K)/P_{KI}$. With pulsed lasers it is easy to reach $P_{KI} \gg A_K$
and $n_a \cong 1$, which implies that a single absorbed photon can be detected! Under these conditions two step resonance ionization spectroscopy is by far the most sensitive method to monitor absorption

spectra. Its experimental drawback is that it requires two lasers.
However, both pulsed dye lasers may be pumped by the same pump laser,
e.g. a nitrogen laser or a frequency doubled or tripled YAG-laser.

With cw-lasers the sensitivity is somewhat lower because of the
above mentioned difficulties. It is, however, still much higher than
that of excitation spectroscopy with photon detection. The lower
limit of detectable absorbed photons is about $n_a \geq 10 - 100$, depending on the molecule under investigation and on the laser powers
available.[10]

V. OODR-SPECTROSCOPY WITH THERMOIONIC ION DETECTORS

If the energy $h(\nu_1 + \nu_2)$ of two photons is not sufficient to
ionize the molecule but allows to reach high lying levels of the
neutral molecule not too far below the ionization limit, a sensitive
detection method can often be utilized which is based on collisional
ionization of these high levels and a charge amplification of the
produced ions. The thermoionic ion detector[11,12] is essentially a
diode with a hot filament, operated under spare charge limited conditions. When a highly excited molecule is formed in the space
charge region, it is readily ionized by collisions with electrons
from the filament. The ion partly compensates the negative space
charge of the electrons and therefore increases the electron flow to
the anode. Since the ion moves very slow compared to the electrons,
it stays for a longer time within the space charge region and a
single ion can increase the electron current by a factor $M = v_{el}/v_{ion}$
which equals the ratio of electron drift velocity to ion drift velocity.

This space charge amplification technique has been successfully
used for the study of highly excited atoms and molecules[13]. Figure
8 illustrates the achievable signal to noise ratio of OODR-signals
obtained with this technique in two step excitation of Na_2-mol-
ecules[14]. The thermoionic diode was in these experiments a heatpipe with a thin heated tungsten wire in the central axis. The
electron current between this wire, which acted as the cathode, and
the walls of the cylindrical heat pipe as anode was measured as a
function of the wavelength λ_2 of the second laser. The first laser
was a single mode argon laser which was stabilized onto the Lamb-dip
of the transition $X^1\Sigma_g^+(v" = 3, J" = 43) \rightarrow B^1\Pi_u(v' = 6, J' = 43)$. A
tunable single mode dye laser was used for further excitation. A
probe laser power of 1 mW was sufficient to record the signal in
Figure 8. The optimum diode voltage was between 0V-4V. The electric
field at the location of the collinear laser beams, a few mm away
from the cathode wire, was sufficiently small to avoid noticeable
Stark shifts of the excited molecular levels.

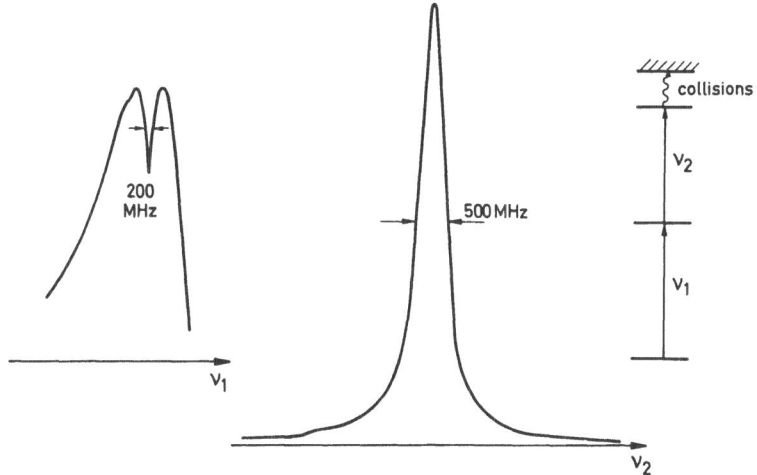

Fig. 8. OODR-signals in a Na_2 heatpipe, monitored with the space
         charge amplification technique. The left signal is obtained
         with 50 mW from a single mode Ar-laser, tunable around λ =
         488 nm, and represents a resonant two-photon excitation with
         the Lamb-dip. The right signal is obtained with the argon
         laser stabilized on the Lamb-dip and a single mode dye laser
         (a few mW) is scanned over a transition from the pumped
         level (v' = 6, J' = 46) in the B-state to a level (v*, J*)
         in a higher Rydberg-state.

VI. MOLECULAR RYDBERG-STATES

Molecular Rydberg-states play an important role in electron-ion
recombination, a process which occurs frequently in the high atmosphere of planets, in the interstellar clouds and in gas discharges
containing molecular species. Although <u>atomic</u> Rydberg-states have
been intensively studied recently[15], not much is known about mol-
<u>ecular</u> Rydberg-states. Their excitation from the ground state by
absorption of a single photon requires sufficiently intense ultraviolet or even vacuum-ultraviolet sources.

Two step excitation by successive absorption of two photons is a
convenient complementary method to populate atomic and molecular
Rydberg states and to unravel the complex and often perturbed structure of transitions between high lying molecular states.

A Rydberg-state (AB)* of a diatomic molecule dissociates into a
ground state atom A and a Rydberg atom B* with a principal quantum
number n. While singly excited <u>atomic</u> Rydberg-states are always
below the ionization limit, this may be different for molecules
(Figure 9). Because of their additional rotational-vibrational
energy, a level (v*,J*) in a molecular Rydberg-state, may be above

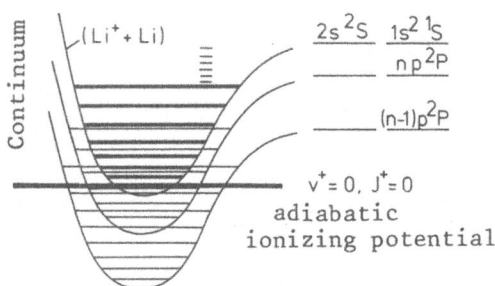

Fig. 9.   Potential curves of molecular Rydberg-states. Rotational
          vibrational levels (v*, J*) of Rydberg-states may autoionize
          if they are above levels (v⁺, J⁺) of the ionic ground state.

the bottom of the ionic ground state potential, which dissociates
into A + B⁺. Such levels can autoionize into an ion (AB)⁺ in the
level (v⁺,J⁺) while an electron is emitted. The final levels (v⁺,J⁺)
which can be reached, depend on the energy of the autoionizing level
(v*,J*) and on certain selection rules.

Since for large principal quantum numbers n the Rydberg-electron
is on the average far away from the molecular core (its mean radius
is about $\langle r \rangle = a_0 \cdot n^2$) it does not significantly contribute to the
binding energy of the core. This implies that the potential curves
$E_n^*(R)$ of the Rydberg-states have a similar form as the ionic curve
$E^+(R)$; they are only shifted by the ionization energy of the electron. Similar to the description of atomic Rydberg-states, these
potential curves $E_n^*(R)$ can be described by a Rydberg-formula

$$E_n^*(R) = E^+(R) - \frac{R_y}{(n - \delta(R))^2}$$
(7)

where R_y represents the Rydberg constant and δ the quantum defect
which allows for the deviation of E_n^* in the actual force field ex-
perienced by the Rydberg-electron, from that in a pure Coulomb-field
where $\delta = 0$. In the molecular case the quadruple moment of the
molecular core and its polarizability determines the actual quantum
defects which depend on the electrons' angular momentum, its principal quantum number n and also slightly on the internuclear separation
R[16,17].

The necessary condition for autoionization of a Rydberg level
$E_n^*(v*,J*)$ into an ionic level $E^+(v^+,J^+)$ can be expressed by the
energy relation

$$E_{vib}^*(v^*) + E_{rot}^* > \Delta E + E_{vib}^+(v^+) + E_{rot}^+(J^+)$$
(8)

where $\Delta E = E^+(R_e) - E_n^*(R_e)$ is the energy separation of the minima of the two potential curves at the equilibrium internuclear distance R_e.

The probability for autoionization depends on the coupling between nuclear motions and electronic energy since part of the rotational-vibrational energy must be converted to the Rydberg electron in order to give it sufficient energy to leave the core. The autoionization process for Rydberg-states of molecules is therefore certainly much slower than that of doubly excited atomic states. In order to estimate its efficiency one has to take into account all other possible decay channels of the excited Rydberg-state $E_n^*(v^*,J^*)$. The autoionization process competes with the radiative decay of the Rydberg levels and often also with possible predissociation. However, the spontaneous transition probability decreases approximately with n^{-3}. For large principal quantum numbers n the radiative decay therefore becomes very slow and in absence of predissociation even small autoionization probabilities may already result in complete ionization of the Rydberg level. Since the ion can be detected with high efficiency, this makes autoionization by two-step excitation a very sensitive variation of resonant two photon ionization, because the transition probability for the second step $k \rightarrow n(v^*,J^*)$ is much higher than that of direct transitions $k \rightarrow I$ into the ionization continuum.

Several groups[18-21] have started experimental work on two step autoionization of diatomic molecules. Our group in Kaiserslautern[22] has chosen the Li_2-molecule because there exist very accurate calculations of ab initio potential curves[24] which can be compared with the experimental results. The experimental procedure is illustrated in Figure 10. The Li_2 molecules are prepared in a supersonic beam which is crossed perpendicularly by two collinear laser beams from two pulsed dye lasers L1 and L2, pumped by the same nitrogen laser. The first laser L1 pumps the Li_2 molecules from a level (v_i'', J_i'') in the $X^1\Sigma_g^+$ ground state into a selected level (v_k', J_k') of the $B^1\Pi_u$-state. Since the molecular constants of both states are known[23] the pump transition $(v_i'',J_i'') \rightarrow (v_k',J_k')$ can be assigned by observing the laser induced fluorescence. When the pump laser L1 is stabilized onto this transition, the second laser L2 is switched on and induces transitions from the intermediate level (v_k',J_k') into higher Rydberg levels (v^*,J^*), which can autoionize. The number of ions is monitored as a function of the wavelength λ_2 of L2, at first with L1 off and then with L1 on. The difference between the two readings gives those ions which are produced by L2 on a transition $(v_k',J_k') \rightarrow (v^*,J^*)$ or by direct photoionization

$$Li_2(v_k', J_k') + h\nu_2 \rightarrow Li_2^+ + e^- \tag{9}$$

In our experiment $2h\nu_2$ was larger than the ionization limit of $Li_2(X^1\Sigma_g)$. The second laser alone can therefore also produce ions by two photon ionization. The probability for this process is, however,

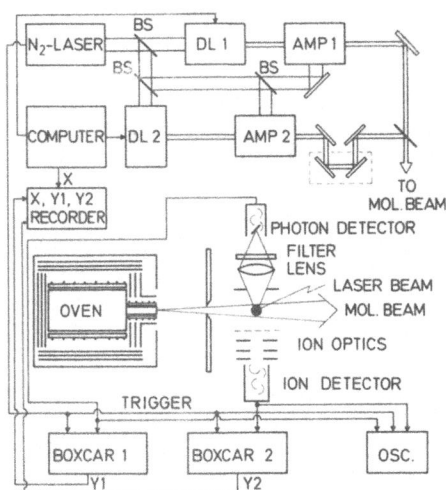

Fig. 10. Experimental arrangement for resonant two-photon ionization
          in a molecular beam. The fluorescence is monitored with a
          photomultiplier, the ions are collected by a small electric
          field and focussed on an open multiplier.

sufficiently large only for resonant two photon ionization when the
photon energy $h\nu_2$ fits to a molecular transition

$$Li_2(X^1\Sigma_g(v'',J'')) + h\nu_2 \rightarrow Li_2 \ B^1\Pi_u(v_2^!,J_2^!) \tag{10}$$

The middle trace (a) in Figure 11 shows the number of ions $N_{ion}(2\nu_2)$
produced with L1 off. The peaks correspond to resonant transitions[10], as could be proved by observation of the corresponding
fluorescence (lowest trace in Figure 11). When the pump laser L1 is
switched on, the upper trace (b) in Figure 11 is recorded. The
difference between the two curves (note that the upper curve is
vertically shifted by an amount indicated by the arrow in order to
separate the two recordings) gives the number of ions produced by one
photon $h\nu_2$ on a transition from a labelled level $(v_k^!,J_k^!)$ in the
intermediate state. The autoionization lines can be seen super-
imposed on a continuum which increases with decreasing wavelength λ_2.
The continuum is attributed to the direct photoionization process (9)
and from its onset at λ_2 = 474.9 nm, measured at different electric
field strengths, the adiabatic ionization potential

$$IP = E(X^2\Sigma_g^+, \ v^+ = 0, \ J^+ = 0) - E(X^1\Sigma_g^+, \ v'' = 0, \ J'' = 0) \tag{11}$$

is determined to be $IP(Li_2^!)$ = 41475 ± 8 cm^{-1}. A comparison between
Figures 11A and 11B shows that with increased strength of the electric field used to extract the ions, some new autoionization lines
appear close above the ionization threshold. They can be attributed

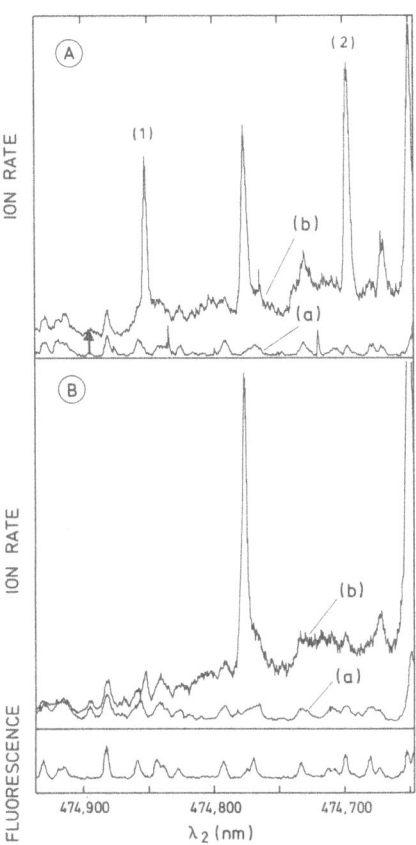

Fig. 11. Section of the two-step photoionization spectrum of Li_2
         around the ionization threshold. Superimposed on a contin-
         uous background due to direct photoionization from the
         intermediate level $B^1\Pi_u$ ($v' = 0$, $J = 6$) the autoionization
         lines can be seen. A: $E = 50$ V/cm, B: $E = 20$ V/cm.

to very high Rydberg levels with $n \cong 50$ and $v^* = 0$ which autoionize
by rotational coupling into $v^+ = 0$. The molecules in these levels
have a large polarizability and the induced dipole moment may enhance
the coupling efficiency.

The linewidth of the autoionization lines is very narrow and
mainly determined by the laser bandwidth of about 3 GHz. In order to
investigate the linewidth in more detail, the bandwidth of L2 was
reduced to 1.5 GHz by inserting an extra etalon into the laser
cavity. The laser linewidth could be determined from the width of
the lines $I_{F1}(\lambda_2)$ induced by L2. Figure 12, which shows two autoionization lines excited at two different field strengths, illustrates that the linewidth increases with increasing electric field.

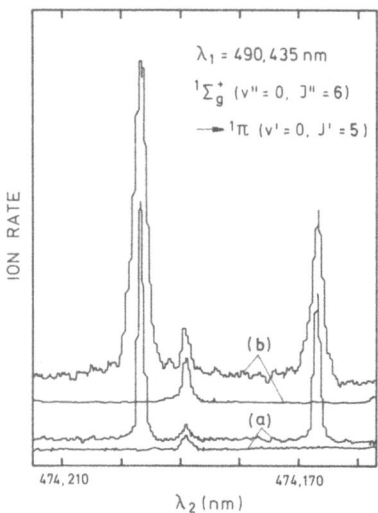

Fig. 12. Linewidth of autoionizing lines at different electric
          fields used to extract the ions. a) E = 20 V/cm, b) E =
          100 V/cm. The lower curves in a) and b) are recorded with
          the pump laser off.

The steps in the line profiles are due to the stepwise scanning of
the laser wavelength. From a comparison of these line profiles with
the fluorescence line profiles excited by the same laser, an upper
limit $\Delta\nu \leq 300$ MHz for the autoionization linewidth at low fields can
be obtained. This implies that the lifetime of the autoionizing
state is at least 0.5 nsec, probably longer[24].

Compared to atomic autoionizing states, where lifetimes below
10^{-12} sec are common, the molecular levels exhibit a much weaker
coupling to the ionization continuum. The reason is that the molecular ionization process takes place through a coupling between
nuclear motion and electronic configurations, whereas in atoms a
direct electronic coupling of the excited electron to the continuum
makes the autoionization probability by far larger.

Similar experiments on autoionization of molecular Rydberg
levels have been performed on Na_2 and K_2 by Martin et al.,[20] and by
Leutwyler et al.,(18,19]. A complete analysis of the autoionization
spectra of all these alkali molecules is, however, still pending
since a full understanding of all perturbations requires very detailed studies of the many molecular states closely below the ionization limit. These may be not only Rydberg-states but also doubly
excited states which dissociate into A* + B*. These states can
perturb the Rydberg-states and congest the spectra.

Several groups have attacked these problems. Bernheim et al.,[25] have performed OODR spectroscopy of Rydberg-states of Li_2 below the ionization limit. The probe transitions were monitored through the fluorescence excited by the probe laser. For a given intermediate level (v', J') of a Σ-state, there are not only P- and R-lines with $\Delta J = \pm 1$ possible for transitions to Σ-Rydberg-states but also simultaneously P, R and Q-lines to π Rydberg-states. Because of perturbations, including l-uncoupling phenomena[26] the assignments of the different lines is not obvious. The authors succeeded, however, to assign several Rydberg series.

Here double-resonance polarization spectroscopy is very helpful because it allows one to distinguish between P, Q or R-transitions[27]. Schawlow et al.,[28,29] have investigated many Rydberg-states of Na_2 by this technique. They used a narrow band pulsed pump laser to label selected intermediate levels (v', J') but a broadband probe laser and recorded many OODR transitions simultaneously on a photographic plate (polarization labelling technique[30]. The technique is also applicable with cw lasers and photoelectric recording with the advantages of sub-Doppler resolution and higher accuracy of wavelength measurements.

Detailed sub-Doppler OODR spectroscopy of BaO has been performed by W. Field and coworkers[31], who used the stepwise excitation scheme of Figure 1b. The resonance transitions k → m of the probe laser were monitored by the decrease of the fluorescence intensity, emitted from level k, as well as by the appearance of a new, ultraviolet fluorescence, emitted from the level m into the electronic groundstate. This method not only allowed a precise characterization of excited electronic states of BaO in the 4 eV region, but also of low lying metastable states.

VII. OODR-SPECTROSCOPY OF DISSOCIATING STATES

Most spectroscopic efforts are directed towards bound states since they give rise to line spectra which allow one to determine molecular constants and potential curves. Much less is known about repulsive states, which are responsible for continuous spectra. They sometimes make themselves conspicuous in discrete spectra by predissociation of bound states which results in broadening of spectral lines. If sufficient predissociating levels can be found, the potential curve of the repulsive states can be sometimes deduced, at least for a certain range of internuclear distances R. However, in many cases it is not clear into which atomic state the repulsive potential curves dissociate.

Collins et al.,[32] used a very elegant variation of OODR-spectroscopy to scan repulsive potential curves of excited states (AB)* and detect the excited atomic states A* into which they dissociate. The method is illustrated in Figure 13. A pulsed tunable dye laser

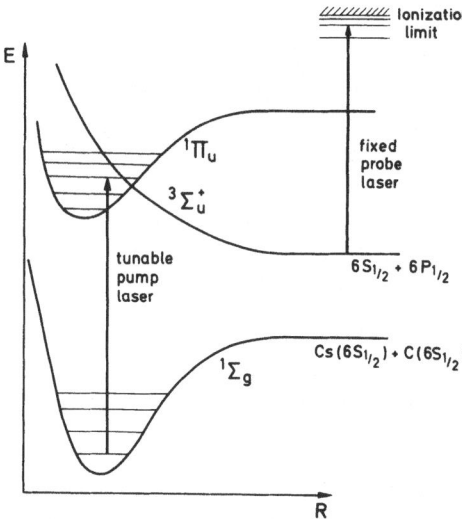

Fig. 13. Potential diagram of Cs_2 for illustration of photolytic double resonance spectroscopy.

is scanned through the molecular spectrum of interest. The pulse from the second dye laser is sent through the sample cell with a short delay. Its wavelength is tuned to the atomic transition between the excited state $\underline{A}^*$ and a higher atomic Rydberg-state A^{**}. When the first laser excites a repulsive molecular state $(AB)^*$ that dissociates into $A^* + B$, the second laser can further excite the atoms A^* produced by photodissociation into the state A^{**} closely below the ionization limit. The sample cell is built as thermoionic detector where the excited atoms A^{**} are ionized by collisions with electrons and are detected with the space charge amplification technique (see Chapter V).

The signal, measured at a selected atomic transition ($\lambda_2 =$ const) as a function of λ_1 yields the probability for transitions from the molecular groundstate to the repulsive state. This allows one to determine the wave function of the repulsive state and its potential curve.

The power of this "photolytic spectroscopy" was demonstrated by applying it to the spectroscopy of repulsive states of Cs_2[33]. Figure 13 shows the corresponding potential diagram. The authors found different repulsive states which dissociate into the atomic Cs states $6P_{1/2}$, $6P_{3/2}$, $5D_{3/2}$ and $5D_{5/2}$ and which overlap in their spectral excitation ranges.

VIII. ACCURATE DETERMINATION OF DISSOCIATION ENERGIES BY OODR

In the Λ-type OODR scheme of Figure 1c pump and probe transitions share a common upper level. Of particular interest are those probe transitions k $\rightarrow$ m that terminate on high vibrational levels m = (v", J") of a lower electronic state or on the continuum above the dissociation limit. Since levels with sufficient energy above the lowest level v" = 0 have negligible populations at thermal equilibrium, the probe wave may even experience gain rather than absorption, provided the upper level k is appreciably populated by the pump transitions. In a proper resonator configuration even laser oscillation can be achieved on these transitions if the total gain overcomes the losses[34]. In this section we briefly discuss applications of the Λ-type OODR scheme to the accurate determination of dissociation energies.

Assume the pump laser excites a high vibrational level v' in an upper electronic state k, less tightly bound than the groundstate. The potential minimum of this state will then be found at a larger internuclear separation R_e than that of the groundstate. When the pump transition starts from a low, thermally population vibrational level in the groundstate i, the maximum Franck-Condon factors occur for those vertical transitions which reach the upper state at the inner limb of the potential curve (see Figure 14). The excited molecule undergoes many vibrations during its spontaneous lifetime and the fluorescence can be in principle emitted at any value of R within the inner and outer turning points R_i and R_o. However, if vertical transitions from the outer turning point can reach the outer part of the groundstate potential, these transitions will have the largest Franck-Condon factors because the vibrating molecule spends a longer time near the outer turning points.

Under such conditions where the upper potential curve is shifted against the lower towards larger internuclear separations, the fluorescence spectrum shows, besides intense transitions between the inner repulsive parts of the potentials around the excitation wavelength, even stronger transitions terminating on very high vibrational levels of the lower state i. Also bound-free continuous fluorescence spectra are found which represent transitions from the upper level (v', J') in state k into continuous states above the dissociation limit[35].

The long wavelength section of such a fluorescence spectrum of Cs_2, as excited by a single mode dye laser on the transition $X^1\Sigma_g^+(v"=0,\ J"=49) \rightarrow D^1\Pi_u^+(v'=50,\ J'=48)$ is shown[36] in Figure 15. From the rotational spacings $\nu(v",J"=J'-1) - \nu(v",J"=J'+1)$ of P and R lines, and from the vibrational spacings $\nu(v"+1,J") - \nu(v",J")$ of lines, terminating on high vibrational levels v" closely below the dissociation limit, the accurate determination of the dissociation energy D_e is possible[37]. For the Cs_2 molecule we obtained from the

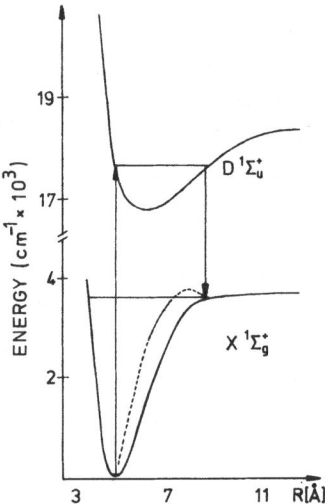

Fig. 14. Schematic potential diagram for the excitation of high
          vibrational levels with subsequent fluorescence to levels
          close to the dissociation limit. The dashed curve re-
          presents the difference potential.

fluorescence spectrum a value $D_e(^1\Sigma_g^+) = 3648 \pm 8 \text{ cm}^{-1}$. From the
structure of the continuous fluorescence spectrum the difference
potential $V_k(R) - V_i(R)$ can be deduced[35].

     The spectral resolution of the fluorescence lines which is
determined by the monochromator, used to disperse the spectrum, is
not high enough to separate the rotational lines for high values of
v'' since their spacings $4B_{v''}(J'+1/2)$ decrease rapidly with increasing
v''. Here the OODR spectroscopy offers much higher resolution and
allows one to resolve rotational spacings very close to the dissociation limit, even when they become narrower than the Doppler-width.

     Figure 16 demonstrates the signal to noise ratio of OODR signals
obtained with polarization spectroscopy of the Cs_2 molecule. The
signal P(72) represents an OODR transition according to the V-scheme
of Figure 1a, where pump and probe transitions share a common lower
level. The line P(74) represents a Λ-type OODR signal according to
Figure 1c where the upper level $J' = 73$ is shared and the probe
transition is due to stimulated emission rather than absorption.
Pump- and probe-waves were travelling antiparallel through the
sample.

     The small signals P(70) and P(76) in Figure 16 are secondary
OODR signals, caused by collisional population transfer within the
lower state from levels $J''=70$ and 76 to the depleted level $J''=72$.

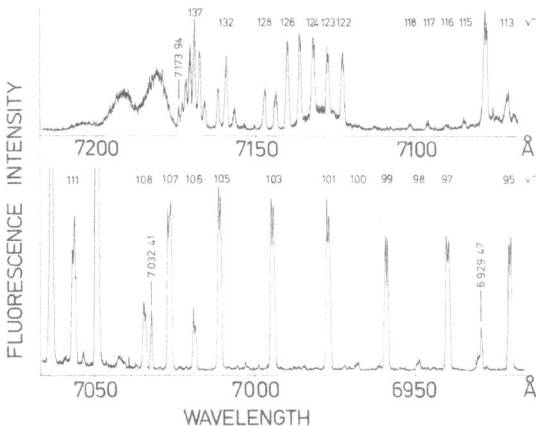

Fig. 15. Laser induced fluorescence spectrum of Cs_2, excited on a
transition $X^1\Sigma_g^+$ (v" = 0, J" = 49) → $D^1\Sigma_u$ (v' = 50, J' = 48)
and terminating on high vibrational levels of the X ground
state. The modulated continuum is due to fluorescence terminating in the continuum above the dissociation limit.

Note, that in homonuclear diatomic molecules collisions can only
induce rotational transitions with even ΔJ values because of symmetry
selection rules.

     High vibrational levels in excited electronic states can be
probed by triple resonance spectroscopy first applied to the Na_2
molecule by R. Field and his coworkers[38]. The technique which is
also called "modulated gain spectroscopy" is illustrated by Figure
17. A single mode dye laser excites the molecules into higher vibrational levels v' in the A- or B-state. With sufficient pump power,
inversions can be reached between v' and high lying levels v" of the
ground state. The resulting optically pumped "dimer-laser" populates
these levels by stimulated emission. A second dye laser is now tuned
to a transition between these high v"-levels and very high v' levels
in the A- or B-state. When the first dye laser is chopped, so will
be the dimer laser intensity and with it the population $N(v_m")$ of the
terminating level, common to dimer laser and second dye laser.

     These experiments can give information on the exact form of the
potentials V(R) in a range of internuclear distances R where the pure
van der Waals potential V(R) = C R^{-6} already fails but quantum-
mechanical calculations, which give optimum results around the potential minimum, are not sufficiently accurate. In particular they can
investigate the influence of spin-orbit coupling in the intermediate
range between Hunds coupling cases (a) and (c) by accurately measuring for example the potential curves of molecular singulet and triplet states dissociating into the same atomic doublet states.

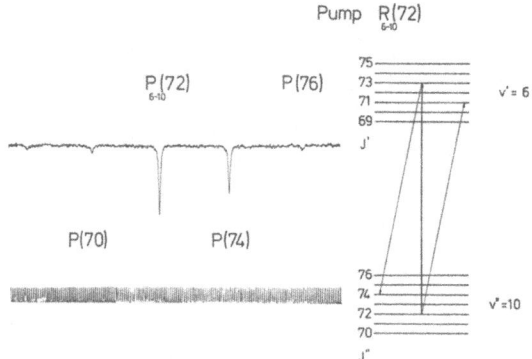

Fig. 16. OODR signals obtained with double resonance polarization
          spectroscopy of Cs_2. The level scheme illustrates the two
          signals P(72) (scheme of Figure 1a) and P(74) (scheme of
          Figure 1c). Pump and probe beams are antiparallel. The
          small signals P(70) and P(76) are collision induced
          "secondary" double resonance signals.

IX. OODR SPECTROSCOPY BELOW THE NATURAL LINEWIDTH

     When the two beams of pump laser and probe laser in the OODR
scheme of Figure 1c travel collinearly instead of antiparallel, a
particular situation arises. The molecule interacts simultaneously
with two copropagating waves which are both in resonance with two
coupled molecular transitions. This may be regarded as a resonant
stimulated Raman scattering process, where the characteristic features of the intermediate level k, such as its lifetime and level-
width, do not enter into the linewidth of the Raman signal. It can
be shown[39] that for this case the linewidth γ_{DR} of the double
resonance signal is given by

$$\gamma_{DR} = \gamma_i + \gamma_m + (1 - \frac{k_{probe}}{k_{pump}}) \gamma_k \qquad (12)$$

where k denotes the wavevector of the probe wave and pump wave. If
both waves travel collinearly, $k_{pump} \parallel k_{probe}$ and the last term in
(12) becomes very small. For $\lambda_{pump} = \lambda_{probe}$ it becomes zero.

     When the levels i and m are rotational-vibrational levels of the
electronic groundstate, they have long radiative lifetimes τ (for
homonuclear molecules τ even becomes infinite). The level widths γ_i
and γ_m are therefore small compared to γ_k. For homonuclear molecules
the width γ_{DR} of the double resonance signal becomes a small fraction
of the width γ_k of the upper level

$$\gamma_{DR} = (1 - \lambda_{pump}/\lambda_{probe}) \gamma_k$$

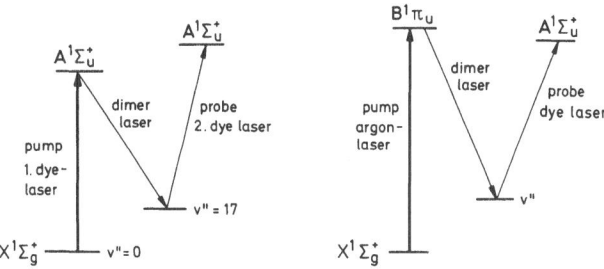

Fig. 17. Excitation schemes for the investigation of high vibrational levels in excited states of Na_2 by modulated gain
spectroscopy.

The Λ-type OODR spectroscopy with copropagating beams therefore
allows a spectral resolution which goes beyond the natural linewidth
γ_k of the optical transitions $i \rightarrow k$ or $k \rightarrow m$.

The technique has been applied by Ezekiel et al.,[40] to the I_2
molecule. The optical transitions, chosen in this case, have however, a very small natural linewidth of approximately 100 KHz.
Because of laser frequency jitter, time of flight broadening and
other broadening effects, it is therefore very hard to obtain linewidths below 100 KHz. The authors could, however, show that the
width γ_{DR} of the double resonance signal was much smaller for copropagating pump- and probe waves than for counter propagating waves.

M. Raab[41] has performed OODR-polarization spectroscopy of the
Cs_2 molecule. Here the spontaneous lifetimes are a few ns and the
natural linewidth is about 20 MHz, which makes it easier to observe
subnatural linewidths. Figure 18 shows two double resonance signals
in the polarization spectrum of Cs_2, which are generated with a fixed
pump laser, stabilized on a transition $X^1\Sigma_g^+$ (v''=4, J''=104) $\rightarrow$ $C^1\Pi_u$
(v'=5, J'=105), and a tunable probe laser. Both laser beams propa-
gate nearly collinearly. The left OODR signal corresponds to the
V-type scheme of Figure 1a, the right signal to the Λ-type scheme of
Figure 1c. Both signals have residual Doppler-widths of about 6 MHz
due to the finite crossing angle of 0.01 rad between the two beams.
Together with the natural linewiths of 20 MHz which is slightly power
broadened to about 25 MHz, this results in a total linewidth of 28
MHz for the OODR signal of the V-type. The Λ-type signal, however,
is only 18 MHz.

This subnatural OODR spectroscopy may be useful to resolve narrow
level spacings of rotational levels of loosely bound molecules close to
the dissociation limit. There are predictions [42] that such molecules have potential minima at very large internuclear distances. The
spacings of rotational levels will then be exceedingly narrow. The problem of rotational predissociation or the occurrence of double minima
potentials in diatomic molecules can be attacked with this technique.

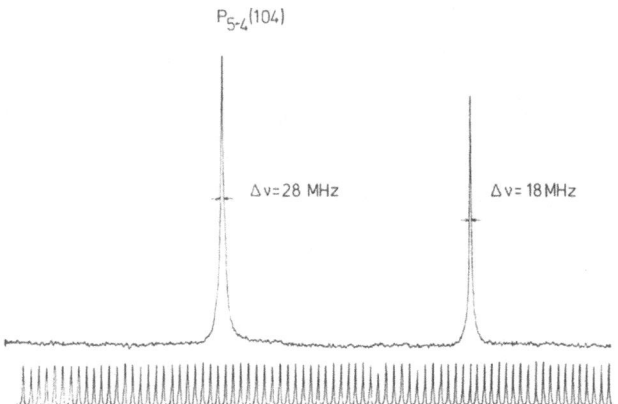

Fig. 18. Two OODR signals with collinear pump and probe beam. Left
          signal: scheme of Figure 1a, right signal: scheme of Figure
          1c. Note the smaller linewidth of the left signal which is
          below the natural linewidth of the optical transition.

Acknowledgements

This lecture intended to give a survey about the basic principles, experimental problems and some applications of OODR spectroscopy. Most of the examples, used to illustrate the techniques, have
been taken from the work of the graduate students in the
Kaiserslautern group, in particular D. Eisel, H. J. Foth, M. Raab, H.
J. Vedder, and H. Weickenmeier. There has been much more work performed, of course, in many other laboratories. Only a small selection of many papers has been mentioned here.

REFERENCES

1. J. L. Hall, Saturated Absorption Spectroscopy, in: "Atomic
    Physics", Vol.3, St. Smith and G. K. Walters, eds., Plenum
    Press, New York, p.615 (1973).
2. M. Raab, G. Höning, W. Demtröder, and C. R. Vidal, High
    resolution laser spectroscopy of Cs_2: II. Doppler-free polar-
    ization spectroscopy of the $C^1\Pi_u \leftarrow X^1\Sigma_g^+$ system, J.Chem.Phys.,
    76:4370 (1982).
3. H. J. Foth, H. J. Vedder, and W. Demtröder, Sub-Doppler Laser
    Spectroscopy of NO_2 in the λ = 592.5 nm Region, J.Mol.
    Spectrosc., 88:109 (1981).
4. D. H. Levy, L. Wharton, and R. E. Smalley, Laser Spectroscopy in
    Supersonic Jets, in: "Chemical and Biochemical Applications
    of Lasers", Vol.II, C. B. Moore, ed., Academic Press, New
    York (1977).
5. G. Herzberg, in: "Molecular Spectra and Molecular Structure",
    Vol.I, p.168, Van Nostrand Reinhold Comp., New York (1950).

6. R. Schmiedl, I. R. Bonilla, F. Paech, and W. Demtröder, Laser Spectroscopy of NO_2 under very high resolution, J.Mol. Spectrosc., 68:236 (1977).

7. W. Demtröder, D. Eisel, H. J. Foth, G. Höning, M. Raab, H. J. Vedder, and D. Zevgolis, Sub-Doppler Laser Spectroscopy of Small Molecules, J.Mol.Structure, 59:291 (1980).

8. E. Gottwald, "Diplomthesis", Fachbereich Physik, Univ. Kaiserslautern (1980).

9. K. Bergmann and E. Gottwald, Effect of Optical Pumping in Two-Step Photoionization of Na_2, Chem.Phys.Lett., 78:515 (1981).

10. M. Becker, "Diplomthesis", Fachbereich Physik, Univ. Kaiserslautern (1982).

11. G. V. Marr and S. R. Wherrett, The Ionization of Caesium Vapor by the Method of Space Charge Amplification, J.Phys.D,At.Mol. Phys., 5:1735 (1972).

12. M. E. Koch and C. B. Collins, Space Charge Ion-Detection of Multiphoton Absorption Phenomena in Lithium Vapor, Phys.Rev., A19:1098 (1979).

13. K. C. Harvey and B. P. Stoicheff, Fine Structure of the n^2D Series in Rubidium near the Ionization Limit, Phys.Rev.Lett., 38:537 (1977).

14. W. Thei , "Diplomthesis", Fachbereich Physik, Univ. Kaiserslautern (1982).

15. A. R. W. McKellar, T. Oka, and B. P. Stoicheff, in: "Laser Spectroscopy V, Part VI", Panel Discussion on Rydberg-States, Springer Series in Opt.Sciences, Vol.30, Berlin, Heidelberg, New York (1981).

16. M. J. Seaton, Quantum Defect Theory, Proc.Phys.Soc., 88:801 (1966).

17. M. Raoult and Ch. Jungen, Calculation of Vibrational Pre-ionization by Multichannel Quantum Defect Theory, J.Chem. Phys., 74:3383 (1981).

18. S. Leutwyler, A. Hermann, L. Wöste, and E. Schuhmacher, Isotope Selective Two-Step Photoionization Studies of K_2 in a Supersonic Molecular Beam, Chem.Phys., 48:253 (1980).

19. S. Leutwyler, M. Hofmann, H. P. Härri, and E. Schuhmacher, The Adiabatic Ionization Potentials of the Alkali Dimers Na_2, NaK and K_2, Chem.Phys.Lett., 77:257 (1981).

20. S. Martin, J. Chevaleyre, S. Valignant, J. P. Perrot, M. Broyer, Chabaud, and A. Hoareau, Autoionizing Rydberg-States of the Na_2 Molecule, Chem.Phys.Lett., 87:235 (1982).

21. D. Eisel and W. Demtröder, Accurate Ionization Potential of Li_2 from Resonant Two-Photon Ionization, Chem.Phys.Lett., 88:481 (1982).

22. D. Eisel, PhD-Thesis, Universität Kaiserslautern (1982).

23. M. M. Hessel and C. R. Vidal, The $B^1\Pi_u \leftarrow X^1\Sigma_g^+$ bandsystem of the $^7Li_2$ molecule, J.Chem.Phys., 70:4439 (1979).

24. P. Botschwina, D. Eisel, W. Müller and W. Demtröder, Chem.Phys., 80:329 (1983).

25. R. A. Bernheim, L. P. Gold, P. B. Kelley, and C. Tomczyk, A Spectroscopic Study of the $E^1\Sigma_g^+$ and $F^1\Sigma_g^+$ States of Li_2 by Pulsed Optical-Optical Double Resonance, J.Chem.Phys., 74: 3249 (1981).

26. U. Fano, Quantum Defect Theory of l-uncoupling in H_2 as an Example of Channel Interaction Treatment, Phys.Rev., A2:353 (1970).

27. R. A. Bernheim, L. P. Gold, P. B. Kelly, T. Tipton, and D. K. Veirs, J.Chem.Phys., 76:57 (1982).

28. N. W. Carlson, A. J. Taylor, and A. L. Schawlow, Identification of Rydberg-States in Na_2 by Two-Step Polarization Labelling, Phys.Rev.Lett., 45:18 (1980).

29. N. W. Carlson, A. J. Taylor, K. M. Jones, and A. L. Schawlow, Two-Step Polarization Labelling Spectroscopy of Excited States of Na_2, Phys.Rev., A24:822 (1981).

30. R. Teets, R. Feinberg, T. W. Hänsch, and A. L. Schawlow, Simplification of Spectra by Polarization Labelling, Phys. Rev.Lett., 37:683 (1976).

31. R. A. Gottscho, P. S. Weiss, and R. W. Field, Sub-Doppler Optical-Optical Double Resonance Spectroscopy of BaO, J.Mol. Spectrosc., 82:283 (1980).

32. C. B. Collins, J. A. Anderson, F. W. Lee, P. A. Vicharelli, D. Popescu, and I. Popescu, Two-Photon Technique for the Dissociative Spectroscopy of Single Molecules, Phys.Rev. Lett., 44:139 (1980).

33. C. B. Collins, F. W. Lee, J. A. Anderson, P. A. Vicharelli, D. Popescu, and I. Popescu, Photolytic Spectroscopy of Simple Molecules I + II, J.Chem.Phys., 74:1053 (1981); 74:1067 (1981).

34. B. Wellegehausen, Optically Pumped cw Dimer Lasers, IEEE J.Quant.Electr., QE-15:1108 (1979).

35. J. Tellinghusen and M. B. Moeller, Chem.Phys., 50:301 (1980).

36. M. Raab, H. Weickenmeier, and W. Demtröder, The Dissociation Energy of the Cesium Dimer, Chem.Phys.Lett., 88:377 (1982).

37. R. J. LeRoy, Energy Levels of a Diatomic near Dissocation, Molecular Spectroscopy, Vol.I, Chem.Soc.Specialist Periodical Report, Chem.Soc.London, pp.113-176 (1973).

38. H. S. Schweda, G. K. Chawla, and R. W. Field, Highly Excited, Normally Inaccessible Vibrational Levels by Sub-Doppler Modulated Gain Spectroscopy: The Na_2 $A^1\Sigma_u^+$ State, Opt.Comm., 42:165 (1982).

39. V. P. Chebotayev, in: "Non Linear Laser Spectroscopy", V. S. Lethokov and V. P. Chebotayev, eds., Springer Series in Opt. Sciences, Vol.4, Springer, Berlin (1979).

40. R. P. Hackel and S. Ezekiel, Observation of Subnatural Linewidths by Two-Step Resonant Scattering in I_2-vapor, Phys.Rev. Lett., 42:1736 (1979).

41. M. Raab, PhD-Thesis, University of Kaiserslautern (1981).

42. W. C. Stwalley, Y. H. Uang, G. Pichler, Phys.Rev.Lett., 41:1164 (1978).

OPTO-ACOUSTICS: OLD IDEA WITH NEW APPLICATIONS

A. C. Tam
IBM Research Laboratory
San Jose
California
USA

INTRODUCTION

The opto-acoustic (OA) or photo-acoustic (PA) effect is the generation of acoustic waves by electromagnetic waves or other types of radiation incident on a sample. This effect was discovered by A. G. Bell in 1880, who observed that audible sound is produced when chopped sunlight is absorbed by a sample. Although the OA effect is very old, there is a great resurgence of interest in it in the past several years, both theoretically and experimentally. This appears to be due to the following reasons:

1) Intense light sources (various lasers and arc lamps) have become more readily available in recent years.
2) Highly-sensitive acoustic detectors (microphones, hydrophones, thin-film piezoelectric detectors, etc) have been developed.
3) OA technique has been shown to be one of the most sensitive spectroscopic technique for gases as well as for condensed samples.
4) OA methods have many unique applications, e.g. spectroscopic studies of opaque or powdered materials, studies of energy conversion processes, and imaging of invisible or subsurface features in solids; furthermore, future "exotic" applications like air-to-ocean communications or cosmic ray detections appear possible.

The fact that the field of OA research and applications is rapidly expanding can be appreciated by noting that several reviews have been published recently, for example, Colles et al.,[1] Kanstad and Nordal[2], Pao[3], Patel and Tam[4], Tam[5], Rosencwaig[6], and Somoano[7]. The aim of the present article is to provide some of the

important points and highlights in OA technique rather than to give a
general review. Both the terms OA and PA are commonly used in the
literature, and they are used interchangeably in this article.

GENERALIZED OPTO-ACOUSTIC EFFECT

The OA effect in a narrow sense is the generation of acoustic
waves due to thermal expansion after optical absorption in a sample,
and this is the usual meaning in the literature. It is however
useful to generalize the OA effect to include other types of acoustic
generation mechanisms by other types of incident energetic beams.
The generalized OA effect with some examples of applications are
shown in Figure 1.

The incident energetic beam can be electromagnetic radiation in
the visible range (as is the usual case in most OA studies) or in
other spectral region from RF to x-rays; it can also be particle
beams like electron, proton, muon, neutrino, etc. Several experiments of acoustic generations by non-optical beams have already been
reported in the literature, e.g. Melcher[8] has reported the acoustic
detection of electron paramagnetic resonance causing RF absorption,
and Learned[9] has discussed the acoustic detection of energetic
particles (cosmic ray muons or neutrinos) in deep oceans. To cause
acoustic generation, the incident beam is usually either pulsed, or
modulated at close to 50% duty cycle; as shown in Figure 1, the
corresponding OA signal is a transient acoustic signal with a well-
defined delay time, or a modulated acoustic signal with a well-
defined phase-shift. Typically, the magnitude of the OA signal
provides a measure of the energy or intensity of the incident beam,
while the delay time or phase-shift provide information on the acoustic propagation time or the de-excitation time in the sample. It is
now obvious that acoustic detection can be advantageously used to
detect many types of energetic beams.

Several other types of OA generation mechanisms besides thermal
expansion are indicated in Figure 1: electrostriction[10], photochemistry[11], molecular dissociation[12], Bubble formation[13], and
breakdown or plasma formation[14], with typically increasing
efficiency of OA generation. For example, typical OA efficiency
(i.e. acoustic energy produced/incident optical energy) for thermal
expansion or electrostriction is less than 10^{-10}, while OA efficiency
for the case of breakdown may be as high as 10^{-1} or more. The many
possible mechanisms imply that many interesting effects can be
studied by OA monitoring, for example, chain reactions in photo-
chemical reaction causes strong acoustic amplifications[12], and weak
laser pulses of 1 mJ energy can generate an acoustic shock wave (via
breakdown in a vapor) that is observable many cm away from the breakdown region[14]. The study and applications of these various OA
generation mechanisms in different systems will be an area of fruitful research.

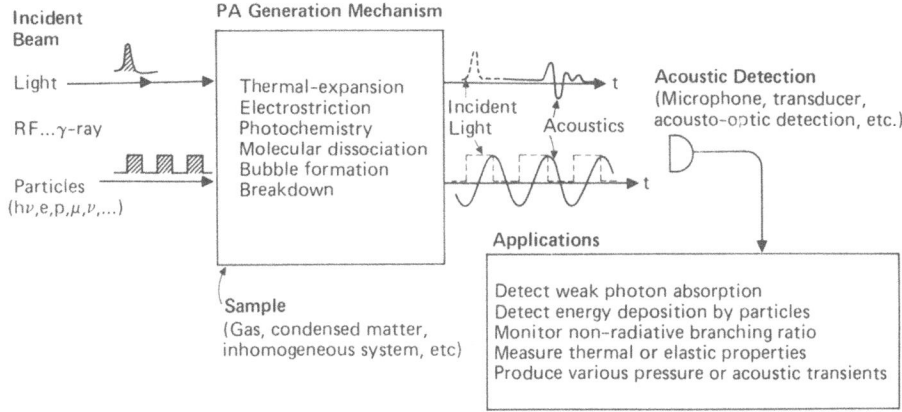

Fig. 1.   The generalized OA effect, showing different possible
           incident beams with pulsed or chopped modulations pro-
           ducing the corresponding acoustic signals.

OA SPECTROSCOPY

The modern interest in the OA spectroscopic detections in gases
appears to begin with the work of Kreuzer[15] in 1971, who reported a
detection sensitivity of 10 ppb of CH_4 in N_2, using a 15 mW HeNe
laser at 3.39 μm for excitation; he also indicated that with a
stronger infrared laser source, it may be possible to detect impurities as low as 10^{-13} by laser OA spectroscopy. This work seems
to be the forerunner of the many later work on OA detection of atmospheric pollutants, for example, Koch and Lahmann[16] have used a
continuous frequency-doubled dye laser of 1 W power to detect SO_2 of
concentrations as low as 10^{-10}. Patel[17] has demonstrated outstand-
ing success of real-time OA detection of low concentrations of NO in
the stratosphere by balloon-flying his OA cell and his laser.

As for the OA spectroscopy of condensed matter, the early work
was performed by Harshbarger and Robin[18], and by Rosencwaig[19],
who simply extended the gas-phase OA spectroscopy technique of
Kreuzer[15], namely, they used a gas-phase microphone to sense the
heating and cooling of a thin gas layer in contact with the solid or
liquid sample which is illuminated by a chopped light beam. This
gas-phase-microphone PA technique for condensed matter relies on the
inefficient thermal diffusion in the sample and in the "coupling"
gas; hence, this technique lacks sensitivity, and is typically useful
only for optical absorptions larger than 1%. Discarding this conventional gas-microphone approach, Patel and Tam[20] have developed a
high-sensitivity pulsed OA technique for condensed matter, with the
use of a tunable pulsed laser for excitation, piezoelectric trans-

ducer in contact with the sample for acoustic monitoring, and a
suitably delayed signal gate and box-car averaging for noise discrimination. Patel and Tam have used this technique in various
liquid and solid samples[4], and concluded that an absorption coefficient as small as $1^{-6}cm^{-1}$ can be measured with their apparatus.
This provides enough sensitivity to obtain new absorption data; for
example, Tam and Patel[21] showed that very accurate absorption
spectra of waters (light and heavy) can be obtained by their pulsed
OA technique.

The above discussions are limited to OA spectroscopy of linear
optical absorptions. Non-linear OA spectroscopy have also been
demonstrated, both for gaseous matter and for condensed matter; these
include Raman-gain spectroscopy[22,23], multi-photon absorption
spectroscopy[24,25], and Doppler-free spectroscopy[26].

OA MONITORING OF DE-EXCITATIONS

After optical excitations of a sample, four de-excitation
branches (with various de-excitation times) are possible: luminescence, photochemistry, photoelectricity, and heat (which may be
generated directly or through energy transfer collisions). Under
sufficiently simple conditions, the OA signal (whose magnitude depend
on the branching ratio for heat production) provides information on
another de-excitation channel. For example, if luminescence and heat
are the only two competing channels, and if the luminescence efficiency can be changed (e.g. by varying the concentrations), the OA
signal magnitude would then be complementary to the luminescence
efficiency, and OA monitoring can yield an absolute value for the
luminescence efficiency without any absolute measurements required[27,28]. Similarly, if photo-carrier generation and heat are
the only two competing channels, the photo-carrier generation
efficiency may be obtained by OA measurements, as performed by
Cahen[29] who studied a solar cell under different external loading
conditions, and by Tam[30] who studied an organic photoconductive
film under different applied electric fields. OA monitoring of
photochemical activities is best illustrated by the work of Cahen et
al.,[31] who showed that the OA signal from "poisoned" chloroplasts
is larger than that from "active" chloroplasts, because there is no
photo-chemical energy conversion in the former.

De-excitation times can sometimes be measured by monitoring the
phase (for modulated excitation at close to 50% duty cycle) or the
time development (for pulsed excitation) of the OA signal. Hunter et
al.,[32] have studied the effect on the phase of the OA signal due to
"fast" heat release and "slow" heat release when optical excitations
produce both singlet and triplet states.

OA PROBING OF PHYSICAL PROPERTIES

   The heating of a sample due to a modulated optical absorption,
and the generation and propagation of acoustic waves depend on the
thermoelastic and physical properties of the sample. Theories for
such processes were first given by White[33]. Hence, by monitoring
the OA signal, we may be able to probe or measure various physical
properties: e.g. acoustic velocites[14,34], thickness[35], crystal-
linity and phase transitions[36], and so on. By focusing the light
beam, some of these physical properties may be measured locally, and
by raster scanning the beam over the sample, new types of "OA
imaging" may be possible, as first demonstrated by Wong et al.,[37]
who showed that subsurface inhomogeneities (not visible with an
optical microscope) in a silicon carbide ceramic sample can be detected by "scanning PA microscopy". The use of a light beam for
excitation is not necessary in these experiments; any other types of
energetic beams (e.g. electrons, ions, etc) which can generate heat
in a sample can be used to provide the same probing as the light
beam; this was shown by Cargill[38], for example, who constructed an
electron-beam-excited acoustic microscope to detect invisible subsurface features in solids.

NON-CONTACT OA MONITORING

   Most OA experiments done so far have been performed by using
microphones or other transducers in contact with or in close prox-
imity with the sample to detect the acoustic signal. These are
contact OA monitoring. However, there are situations when non-
contact OA monitoring may be necessary or prefered, for example,
inaccessible sample (e.g. sample in a vacuum chamber), hostile
environments, or samples that cannot be contaminated. Several ways
for non-contact photo-thermal detections have been demonstrated in
the literature. For example, Nordal and Kanstad[39] have developed
"photo-thermal radiometry", which relies on the detection of the
increased black-body radiation from a sample after optical absorption
and thermal de-excitation; such a technique has been shown to be
useful for spectroscopy and for subsurface imaging. Also, the
photo-thermal heating produces a thermal refractive index gradient
which diffuses from the region of optical absorption; this thermal
refractive index causes "thermal lensing effects"[40], and "probe-
beam refraction"[41] effects, both of which are shown to be useful
for spectroscopy and the other OA applications except for determining
the acoustic properties of the sample. In other words, these non-
contact photo-thermal techniques do not provide information on the
acoustic properties, because no acoustic propagations are involved.

   Tam et al.,[14] first demonstrated a simple method for non-
contact OA monitoring, where the propagating acoustic pulse generated
by an optical pulse is detected by a continuous probe beam; the

arrival of the acoustic pulse at the probe is indicated by a transient probe-beam deflection. This non-contact OA monitoring method can be used for all the OA applications mentioned above, including obtaining acoustic properties. For example, Zapka and Tam have used this method for non-contact measurement of ultrasonic velocities at various temperatures in corrosive liquids[34], of flow-velocities and temperature in a flowing gas[42], and of flame temperature profiles[43]. This last work clearly manifests the advantage of non-contact OA monitoring, namely extending OA techniques into highly hostile environments.

THEORIES OF OA GENERATION AND DETECTION

     We have mainly discussed experimental investigations of OA generations and applications so far. However, there has also been much theoretical work published. Early theoretical investigations of acoustic pulse generation by transient surface heating have been reported by White[33], Gourney[44], Hu[45], and others. Theories on the interpretation of PA signals in condensed matter using gas-phase microphone for detection have been given by Rosencwaig and Gersho[46], McDonald and Wetsel[47], and others. Recently, there is much interest in the quantitative understanding of the shape of acoustic pulse produced by various conditions of optical excitation; i.e. variations of the temporal or spatial profiles of the incident optical beam. There appears to be several reasons for these theoretical work:

1) to optimize OA detection for weak absorbance detection[48,49],
2) to understand OA generations by various nonlinear effects[50],
3) to optimize acoustic detection methods for energetic particles traversing large masses[9], and
4) to distinguish the acoustic pulse shape produced by thermal-elastic effects from that produced by other effects, e.g., electrostriction[51].

These theoretical investigations have improved our understanding of the OA pulse shape significantly, although a detailed and precise comparison with experimental OA pulse shapes is still lacking. It is expected that quantitative understanding of OA pulse shape and directionality would lead to important future applications, including air-to-submarine communications, industrial material testing and quality controls, geo-physical tests and oil explorations, and so on.

REFERENCES

1. M. J. Colles, N. R. Geddes, and E. Mehdizadeh, Contemp.Phys., 20:11 (1979).
2. S. O. Kanstad and P. E. Nordal, Appl.Surf.Sci., 5:286 (1980).

3. Y. H. Pao, "Opto-acoustic Spectroscopy and Detection", Academic Press, New York, (1977).
4. C. K. N. Patel and A. C. Tam, Rev.Mod.Phys., 53:517 (1981).
5. A. C. Tam, in: "Ultra-sensitive Spectroscopic Techniques", D. Kliger, ed., Academic Press, New York, (1983).
6. A. Rosencwaig, in: "Photoacoustics and Photoacoustic Spectroscopy", John Wiley, New York (1980).
7. R. B. Somoano, Angew.Chem.Int.Ed.Eng., 17:238 (1978).
8. R. L. Melcher, Appl.Phys.Lett., 37:895 (1980).
9. J. G. Learned, Phys.Rev., (1979).
10. S. R. J. Brueck, H. Kildal, and L. J. Belanger, Opt.Communic., 34:199 (1980).
11. R. R. Chance and M. L. Shand, J.Chem.Phys., 72:948 (1980).
12. G. J. Diebold and J. S. Hayden, Chem.Phys., 49:429 (1980).
13. G. A. Askaryan, A. M. Prokhorov, G. F. Chanturiya, and G. P. Shipulo, Sov.Phys.JETP, 17:1463 (1963).
14. A. C. Tam, W. Zapka, K. Chiang, and W. Imaino, Appl.Opt., 21:69 (1982).
15. L. B. Kreuzer, J.Appl.Phys., 42:2934 (1971).
16. K. P. Koch and W. Lahmann, Appl.Phys.Lett., 32:289 (1978).
17. C. K. N. Patel, Science, 202:157 (1978).
18. W. R. Harshbarger and M. B. Robin, Acc.Chem Res., 6:328 (1973).
19. A. Rosencwaig, Opt.Communic., 7:305 (1973).
20. C. K. N. Patel and A. C. Tam, Appl.Phys.Lett., 34:467 (1979).
21. A. C. Tam and C. K. N. Patel, Appl.Opt., 18:3348 (1979).
22. J. J. Barrett and M. J. Berry, Appl.Phys.Lett., 34:144 (1979).
23. C. K. N. Patel and A. C. Tam, Appl.Phys.Lett., 34:760 (1979).
24. D. M. Cox, Opt.Communic., 24:336 (1978).
25. A. C. Tam and C. K. N. Patel, Nature, 280:304 (1979).
26. E. E. Marinero and M. Stuke, Opt.Communic., 30:349 (1979).
27. M. J. Adams, J. G. Highfield, and G. F. Kirkbright, Anal.Chem., 49:1850 (1977).
28. R. S. Quimby and W. M. Yen, Opt.Lett., 3:181 (1978).
29. D. Cahen, Appl.Phys.Lett., 33:810 (1978).
30. A. C. Tam, Appl.Phys.Lett., 37:978 (1980).
31. D. Cahen, S. Malkin, and E. J. Lerner, FEBS Lett., 91:339 (1978).
32. T. F. Hunter, D. Runbles, and M. G. Stock, J.Chem.Soc.Faraday II, 70:1010 (1974).
33. R. M. White, J.Appl.Phys., 34:3559 (1963).
34. W. Zapka and A. C. Tam, Appl.Phys.Lett., 40:310 (1982).
35. A. C. Tam and H. Coufal, Appl.Phys.Lett., submitted.
36. J. F. McClelland and R. N. Kniseley, Appl.Phys.Lett., 35:121 (1979).
37. Y. H. Wong, R. L. Thomas, and G. F. Hawkins, Appl.Phys.Lett., 32:538 (1978).
38. C. S. Cargill, Nature, 286:691 (1980).
39. P. E. Nordal and S. O. Kanstad, Physica Scripta, 20:659 (1979).
40. R. L. Swofford, M. E. Long, and A. C. Albrecht, J.Chem.Phys., 65:179 (1976).

41. A. C. Boccara, D. Fournier, W. Jackson, and N. M. Amer,
     Opt.Lett., 5:377 (1980).
42. W. Zapka and A. C. Tam, Appl.Phys.Lett., 40:1015 (1982).
43. W. Zapka, P. Pokrowsky, and A. C. Tam, Opt.Lett., October
     (1982).
44. L. S. Gourney, J. Acoust.Soc.Am., 40:1322 (1966).
45. C -L. Hu, J.Acoust.Soc.Am., 46:728 (1969).
46. A. Rosencwaig and A. Gersho, J.Appl.Phys., 47:64 (1976).
47. F. A. McDonald and G. C. Wetsel, Jr., J.Appl.Phys., 49:2313
     (1978).
48. A. Atalar, Appl.Opt., 19:3204 (1980).
49. G. Liu, Appl.Opt., 21:955 (1982).
50. L. M. Lyamshev and K. A. Naugol'nykh, Sov.Phys.Acoust., 27:357
     (1981).
51. H. M. Lai and K. Young, J.Acoust.Soc.Am., 72:2000 (1982).

PHOTOACOUSTIC ANALYSIS IN CONDENSED MATTER

F. Scudieri

II Università di Roma
Roma
Italy

INTRODUCTION

The photoacoustic (P.A.) effect, sometimes known as the opto-acoustic effect, was discovered by A. G. Bell [1] in 1880, and also observed in 1881 by W. Rontgen [2] and J. Tyndall [3]. Although such effect has been known for a long time no developments in studying physical properties of materials have been reported until 1973. This delay in the improvement of the P. A. techniques is due to the fact that lasers are used as sources in such techniques, even if incoherent broad band light sources have shown to be useful sources for optical excitation.

The basic physical process in the P. A. effect is that, when an intensity modulated light beam is absorbed by a given medium, part of such energy is converted into heat. Such heat can be detected either by measuring, by means of a sensitive microphone, the time dependent pressure fluctuation induced in a coupling gas, or by measuring, by means of a piezoelectric transducer, the thermal stresses or strains in the absorbing medium. The revived great interest towards this effect shown in the 70's is connected with the ability of the technique to measure optical absorption coefficient in an extremely wide range of frequencies, to give information on thermal properties of a material and to be a unique tool for investigation of non radiative processes in matter.

In the following the main P. A. techniques will be discussed and the applications to condensed matter analysis will be presented.

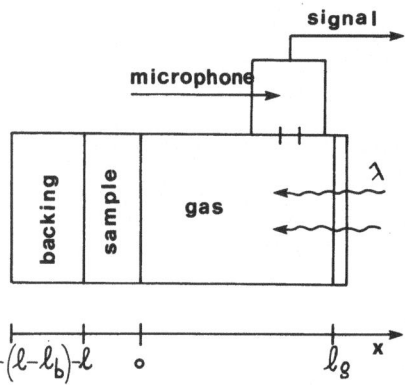

Fig. 1. Geometry for gas-microphone technique.

PHOTOACOUSTIC TECHNIQUES

Gas-microphone P. A.

In the present paragraph the unidimensional Rosencwaig and
Gersho [4] theory will be reviewed. Let us consider an absorbing
sample in contact with a backing support and an inert gas inside a
closed cell (Figure 1); a microphone detects the pressure change of
the gas column. The overall length of the cell is much smaller than
the wavelength of the acoustic perturbation so as to exclude all
resonance processes (resonant cells are extensively used to increase
the signal to noise ratio, but the experimental conditions are particularly critical). An intensity modulated monochromatic radiation

$$I = \frac{1}{2} I_0 (1 + \cos \omega t)$$

passing through a transparent window impinges onto the sample surface
in contact with the gas. The thermal energy density produced at x
is given by $\frac{1}{2} \beta I_0 e^{\beta x}(1 + \cos \omega t)$, where β is the absorption coefficient
at the incident wavelength λ.

In connection with the thermal diffusion problem inside the
cell, we can define the following parameters: k_j thermal conductivity,
ρ_j specific mass, c_j specific heat, $\alpha_j = \dfrac{k_j}{\rho_j c_j}$ thermal diffusivity,
$a_j = \sqrt{\dfrac{\omega}{2\alpha_j}}$ thermal diffusion coefficient, $\mu_j = a_j^{-1}$ thermal diffusion
length where j = g,s,b indicates gas, sample and backing support
respectively. The thermal diffusion equations in the steady state
approximation are

$$\frac{\partial^2 \Phi}{\partial x^2} = \alpha_j^{-1} \frac{\partial \Phi}{\partial t} - \frac{\beta I_0 \eta}{2k_s} e^{\beta x}(1+e^{i\omega t}); \quad [\text{in the sample } -\ell \leqslant x \leqslant 0]$$

$$\frac{\partial^2 \Phi}{\partial x^2} = \alpha_b^{-1} \frac{\partial \Phi}{\partial t}; \quad [\text{in the backing } -(\ell+\ell_b) < x < -\ell]$$

$$\frac{\partial^2 \Phi}{\partial x^2} = \alpha_g^{-1} \frac{\partial \Phi}{\partial t}; \quad [\text{in the gas } 0 \leqslant x \leqslant \ell_g]$$

(1)

where the gas and backing support are supposed to be quite transparent to the impinging radiation, and η is the conversion efficiency of the radiation into heat. The boundary conditions are assumed to be as follows:

i)   temperature and heat flux are continuous functions at $x = 0$ and $x = -\ell$;

ii)   temperature values at $x = \ell_g$ and $x = -(\ell+\ell_g)$ are equal to room temperature;

iii) all effects connected with mass convection in the gas can be neglected.

With the mentioned boundary conditions the solution of Eqs. (1) is

$$\Phi(x,t) = \begin{cases} \ell_b^{-1}(x+\ell+\ell_b)W_0 + We^{\sigma_b(x+\ell) + i\omega t}; & -(\ell+\ell_b) \leqslant x \leqslant -\ell \\ e_1 + e_2 + de^{\beta x} + [Ue^{\sigma_s x} + Ve^{-\sigma_s x} - Ee^{\beta x}]e^{i\omega t}; & -\ell \leqslant x \leqslant 0 \\ (1-x\ell_g^{-1})\theta_0 + \theta e^{(-\sigma_g x + i\omega t)}; & 0 \leqslant x \leqslant \ell_g \end{cases}$$

(2)

where W, U, V, E, θ are complex constants; e_1, e_2, d, W_0, θ_0 are real constants, and

$$\sigma_j = (1+i)a_j; \quad a_j = \sqrt{\frac{\omega}{2\alpha_j}}.$$

The expression for the temperature complex amplitude at $x = 0$ is given by

$$\vartheta = \frac{\beta I_0}{2k_s(\beta^2-\sigma_g^2)}$$

$$\times \frac{(r-1)(b+1)e^{\sigma_s\ell}-(r+1)(b-1)e^{-\sigma_s\ell}+2(b-r)e^{-\beta\ell}}{(g+1)(b+1)e^{\sigma_s\ell}-(g-1)(b-1)e^{-\sigma_s\ell}}$$

(3)

where

$$b = \frac{k_b a_b}{k_s a_s}; \quad g = \frac{k_g a_g}{k_s a_s}; \quad r = \frac{(1-i)\beta}{2a_s}; \quad \sigma_s = (1+i)a_s.$$

Therefore the $\Phi(x,t)$ expression is given by

$$\Phi(x,t) = \vartheta\, e^{-\sigma_g x + i\omega t}$$

whose real part represents the temperature increase

$$T(x,t) = e^{-a_g x}\left[\vartheta_1 \cos(\omega t - a_g x) - \vartheta_2 \sin(\omega t - a_g x)\right]. \tag{4}$$

ϑ_1 and ϑ_2 are the real and imaginary part of ϑ respectively. Due to the spatial behavior of $T(x,t)$ in Eq. (4), practically only in the thickness $2\pi\mu_g = 2\pi/a_g$ of the gas, temperature change is different from zero; that is, only such a thickness follows the temperature variation of the sample. Such a gas layer, whose mean temperature value $\bar\Phi$ is given by

$$\bar\Phi(t) = \frac{1}{2\pi\mu_g}\int_0^{2\pi\mu_g} \vartheta\, e^{-\sigma_g x + i\omega t}\, dx = \frac{\vartheta}{2\pi\sqrt{2}}\, e^{i\omega t - i\frac{\pi}{4}}, \tag{5}$$

behaves, through to the periodic heating of the sample, as a piston on the cell gas giving rise to a pressure change. Considering that the piston displacement is given by

$$\delta_x(t) = 2\pi\mu_g\, \frac{\overline{\Phi(t)}}{T_0} = \frac{\mu_g \vartheta}{I_0\sqrt{2}}\, \exp\left[i\left(\omega t - \frac{\pi}{4}\right)\right]$$

where T_0 is the constant component of the sample temperature and $\upsilon = \upsilon_0\, \frac{T}{T_0}$, and that the gas behavior is adiabatic ($p\upsilon^\gamma = \text{const}$), we obtain for the gas pressure changes in the cell

$$\delta p(t) = \frac{\delta p_0}{\ell_g}\, \delta_x(t) = \frac{\delta p_0 \mu_g \vartheta}{T_0 \ell_g \sqrt{2}}\, e^{i\left(\omega t - \frac{\pi}{4}\right)} = Q e^{i\left(\omega t - \frac{\pi}{4}\right)}. \tag{6}$$

The pressure change $\Delta P(t)$ measured by the microphone is given by the real part of Eq. (6), which by writing

$$Q = Q_1 + iQ_2 = q e^{-i\psi} \tag{7}$$

is

$$\Delta P(t) = Q_1 \cos\left(\omega t - \frac{\pi}{4}\right) - Q_2 \sin\left(\omega t - \frac{\pi}{4}\right) = q\cos\left(\omega t - \psi - \frac{\pi}{4}\right), \tag{8}$$

where Q_1, Q_2, q, ψ are respectively the real part, the imaginary part, the modulus and phase of Q. Eq. (7) together with Eqs. (6) and (3) allows the determination of the needed modulus and phase of the signal detected by the microphone.

The previous result has been obtained in the hypothesis of instantaneous non radiative process; if such a process is characterized by a decay time τ, the Eq. (8) must be multiplied by the factor $\frac{1}{1+i\omega t}$ [5]. The same authors Mandelis et al. [5] have developed a theory for bilayered samples that has been applied to the study of ions implanted semi-conductors [6].

For a better comprehension of the Rosencwaig-Gersho theory the following different cases can be condensed. Assuming $g < b$ and $b \simeq 1$, $\mu\beta = \frac{1}{\beta}$ the optical penetration length and $Y = \gamma P_0 I_0 / (2\sqrt{2} \ell_g T_0)$, we have that, in the case of:

a)   Optically transparent solid ($\mu\beta > \ell$):
     a_1) For thermally thin solid ($\mu_s \gg \ell$, $\mu_s > \mu_\beta$),

$$Q \simeq \frac{(1-i)\beta\ell}{2a_g} \left(\frac{\mu_b}{k_b}\right) Y. \qquad (9)$$

The acoustic signal is proportional to $\beta\ell$ and to ω^{-1} (because μ_b/a_g is proportional to ω^{-1}). Only the thermal properties of the support are involved.

     a_2) For thermally thin solid ($\mu_s > \ell$, $\mu_s < \mu_\beta$),

$$Q \simeq \frac{(1-i)\beta\ell}{2a_g} \left(\frac{\mu_b}{k_b}\right) Y \qquad (10)$$

which is the same as Eq. (9).

     a_3) For thermally thick solid ($\mu_s < \ell$, $\mu_s \ll \mu_\beta$),

$$Q \simeq -i \frac{\beta\mu_s}{2a_g} \left(\frac{\mu_s}{k_s}\right) Y. \qquad (11)$$

The acoustic signal is proportional to $\beta\mu_s$; i.e., only the radiation absorbed in the thermal diffusion length contributes to the signal. The chopping frequency dependence is $\omega^{-3/2}$; in such conditions, by varying ω it is possible to obtain the optic absorption coefficient profile.

b)   In the case of optically opaque solid ($\mu_\beta \ll \ell$):
     b_1) Thermally thin solid ($\mu_s \gg \ell$, $\mu_s \gg \mu_\beta$),

$$Q \simeq \frac{(1-i)}{2a_g} \left(\frac{\mu_b}{k_b}\right) Y. \qquad (12)$$

Q is independent from β and only the thermal properties of the support are involved.

     b_2) For thermally thick solid ($\mu_s < \ell$, $\mu_s > \mu_\beta$)

$$Q \simeq \frac{(1-i)}{2a_g} \left(\frac{\mu_s}{k_s}\right) Y. \qquad (13)$$

b_3) For thermally thick solid ($\mu_s \ll \ell$, $\mu_s > \mu_\beta$),

$$Q \simeq - \frac{i\beta\mu_s}{2a_g} \left(\frac{\mu_s}{k_s}\right) Y. \tag{14}$$

The radiation absorbed in the thermal diffusion length contributes
to the signal. The sample is not photoacoustically opaque as in the
a_3) case.

A typical experimental set-up for P. A. spectroscopy is shown
in Figure 2. The light source can be a Xe lamp ($\sim$ 1000 W), a cw dye
laser in the visible wavelength range, a Globar or spin-flip Raman
laser or a parametric amplifier for IR. The intensity modulator is
electromechanical in the low frequency range (0.5 ÷ 3000 Hz), and
acoustic-optic for higher frequency range. For the cell, the
following factors must be taken into consideration:

- good acoustic insulation must be provided;
- the noise due to the interaction of the impinging radiation
with the empty cell (windows, walls, microphone) must be kept low;
- acoustic resonance must be avoided;
- the sample should have a suitable shape.

The unidimensional Rosencwaig-Gersho theory is still valid, as
pointed out by several authors [7-9], when the thermal diffusion
length in the gas is smaller than the lateral dimension of the cell.

PIEZOELECTRIC P. A.

The P. A. technique which we have just discussed shows some
disadvantages; i.e., it is difficult to rise a microphone at very
low temperature and the absorption coefficient which is measured is

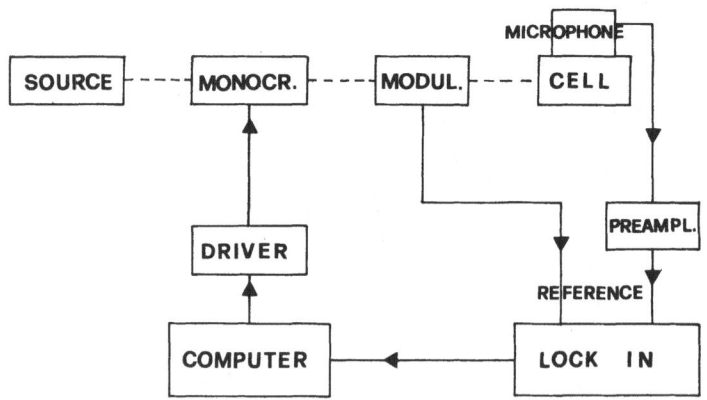

Fig. 2. Block diagram for P. A. spectroscopy.

the one which involves in the sample thermal length. A technique that overcomes such difficulties is the one which uses a piezoelectric transducer attached directly to the sample [10-13]. Jackson and Amer [13] have developed the theory for piezoelectric P. A. of condensed matter samples. They consider the case of a thin ring of a piezoelectric ceramic axially poled, attached on the sample surface by means of an adhesive (Figure 3).

If a light beam is incident on the absorbing solid sample the consequent temperature increase gives rise to a deformation of sample itself. In fact, the expansion of the illuminated region causes an increase of the size of both sides of the sample and the non-uniform heat generation inside the solid gives rise to bending of the sample; such induced strains are detected by the PZT detector. The authors assume that the sample responds as if its boundaries were free from stresses; they neglect the inertial term (that is true when operating at low chopping frequencies) and the heat caused by the stress wave. In such conditions the temperature distribution in the sample can be obtained from the equation:

$$\nabla^2 T - \frac{1}{\lambda} \frac{\partial T}{\partial t} = - \frac{Q(r,t)}{k} \qquad (15)$$

with the boundary conditions

$$k\nabla T\big|_{z=0} = k_1 T\big|_{z=0}; \ k\nabla T\big|_{z=\ell} = - k_2 T\big|_{z=\ell}, \qquad (16)$$

where T is the temperature, k is the thermal conductivity, $\lambda = k/(\rho C)$ is the thermal diffusivity, ρ is the density, C is the specific heat at constant volume, k_1 and k_2 are the surface conductivities of the front and rear surfaces respectively, and Q(r,t) is the power density deposited by the laser beams.

Assuming that the laser beam is square wave intensity modulated and with a gaussian profile, and performing a lock-in detection, it

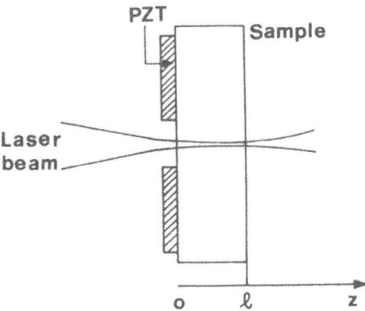

Fig. 3. PZT detection of P. A. signal.

is possible to write for Q(r,t) the following expression:

$$Q(r,t) = \frac{2P_0(1-R)}{\pi^2 a^2} \, \alpha e^{-\frac{r^2}{a^2}} e^{-\alpha z} \cos(\omega t), \tag{17}$$

where P_0 is the incident power, R the reflectivity, a the laser beam radius, α the optical absorption coefficient, ω the angular modulation frequency. With such a choice for Q(r,t) it is possible to determine T(z,r,t).

   The next step is to compute the induced strain in the sample, and it can be obtained by solving the following set of equations:

$$\nabla^2 u_r - \frac{1}{r^2} u_r + \frac{\partial e}{\partial r} - \frac{2(1+\nu)}{1-2\nu} \, a_t \, \frac{\partial T}{\partial r} = 0,$$

$$\nabla^2 u_z + \frac{1}{1-2\nu} \frac{\partial e}{\partial r} - \frac{2(1+\nu)}{1-2\nu} \, a_t \, \frac{\partial T}{\partial r} = 0 \tag{18}$$

with

$$e = \frac{\partial u_r}{\partial r} + \frac{1}{r} u_r + \frac{\partial u_z}{\partial z},$$

$$\nabla^2 = \frac{\partial^2}{\partial r^2} + \frac{1}{r} \frac{\partial}{\partial r} + \frac{\partial^2}{\partial z^2}$$

where u_i is the displacement in the ith direction, ν is Poisson's ratio, a_t is the linear expansion coefficient of the sample. The boundary conditions are that the sample is stress free at z = 0 and z = ℓ; such a situation corresponds to the case of thin piezoelectric detector. If the strains and stresses induced in the material are also computed, it is then possible to determine the corresponding stress detected by the transducer. For a piezoelectric material we have:

$$\sigma_{ij} = C^E_{ijk\ell} u_{k\ell} - e_{kij} E_k,$$

$$D_i = e_{ik\ell} u_{k\ell} + \varepsilon^s_{ik} E_k \tag{19}$$

where σ_{ij} is the stress, u_{ij} is the strain, E_k is the electric field, D_i is the displacement, $C^E_{ijk\ell}$ is the compliance, $e_{ik\ell}$ is the piezo-electric constant and ε^s_{ik} the dielectric constant of the detector. By using Eqs. (19) we can write the final expression for the potential V at the electrodes of the transducer as:

$$V \simeq \frac{e^P_{31} L a_t}{\varepsilon^P_{33} A} (1+\nu) \left[<T_0> + (z - \frac{1}{2}) <\tau> \right]_{z=0,\ell}, \tag{20}$$

where

$$T_0 = \frac{1}{e} \int_0^\ell T(xy\bar{z},t)d\bar{z},$$

$$\tau = \frac{12}{\ell^3} \int_0^\ell (\bar{z} - \frac{1}{2}\ell)T(xy\bar{z},t)d\bar{z}$$

are the z-averaged temperature and averaged gradient, L and A are the thickness and area of the transducer. The first term in Eq. (20) is due to the in-plane displacement connected to the average temperature T_0, the second term is due to the sample buckling connected to the average temperature gradient τ. Eq. (20) can, in some cases, be simplified:

a)   For thermally and optically thick sample with transducer placed at $z = \ell$ (i.e., $\ell \gg (\frac{2\lambda}{\omega})^{1/2}$; $\ell \gg \frac{1}{\alpha}$),

$$V \simeq - \frac{4e_{31}^P L}{\varepsilon_{33}^P A\pi} (1+\nu) \frac{P_0(1-R)a_t}{i\omega\ell(\rho C)_{sample}}. \tag{21}$$

b)   For thermally thin and optically thick sample with transducer placed at $z = \ell$ (i.e., $(\frac{2\lambda}{\omega})^{1/2} \gg \ell \gg \frac{1}{\alpha}$),

$$V \simeq \frac{2\ell_{31}^P L}{\varepsilon_{33}^P A\pi} (1+\nu) \frac{P_0(1-R)a_t}{i\omega\ell(\rho C)_{sample}}. \tag{22}$$

c)   For thermally thick but optically thin sample (i.e., $\frac{1}{\alpha} \gg \ell \gg \gg (\frac{2\lambda}{\omega})^{1/2}$),

$$V \simeq \frac{2e_{31}^P L}{\varepsilon_{33}^P A\pi} (1+\nu) \frac{P_0(1-R)a_t}{i\omega\ell(\rho C)_{sample}} (1-e^{-\alpha\ell} \pm$$

$$\pm \frac{\sigma}{\ell\alpha}[(1-\frac{\ell\alpha}{2}) - e^{-\alpha\ell}(1+\frac{\ell\alpha}{2})]), \tag{23}$$

where the negative sign applies in the case where the transducer is away ($z = \ell$) from the laser beam and the positive sign ($z = 0$) in the opposite case.

   The previous theoretical results have been tested by the authors for glass and a few metals. Also, the results are appreciable with the drastic hypothesis imposed in the theory; in particular, the alteration imposed by the transducer on the sample expansion and the pyroelectric effect in the piezoelectric material are neglected. A limitation to the detector performance is imposed by the scattered light that can reach the transducer. Jackson and Amer [13] do not give any limiting value for the optical absorption coefficient.

Such values are determined by Hordvik and Schlossberg [11] to be in the 10^{-6} cm^{-1} range using laser powers of about 1W. The piezo-electric P. A. technique has been used by Mordvik and Skolnik [14] to determine both bulk and surface losses in transparent solids.

PULSED OPTOACOUSTIC ANALYSIS

According to some authors [15] the gas-microphone P. A. technique must be considered as a photothermal technique and not as a photo-acoustic one, because the detected signal is correlated to the fluctuating temperature of the sample surface exposed to the gas, whereas all possible acoustic signals generated in the sample are generally negligible. A technique where the detected signal is the acoustic one due to the adiabatic isobaric expansion of the sample region illuminated by a light pulse, is used by Patel and Tam [16].

Let us consider a light pulse, whose duration is $\tau_p = 2\tau$, illuminating a cylindrical region of radius R and length ℓ of a weakly absorbing liquid; i.e., $\alpha\ell \ll 1$ where α is the optical absorption coefficient of the medium at the light wavelength λ (Figure 4). If a pressure detector, such as a piezoelectric one, is placed at a distance r, and assuming that the pulse duration time τ_p is much greater than the non radiative relaxation time τ_{NR}, the time τ_a for an acoustic pulse to travel across the illuminated region (i.e., the local pressure equilibrium is maintained) and the response time of the transducer τ_t is

$$\tau_p \gg \tau_{NR}; \quad \tau_p \gg \tau_a = \frac{2R}{\upsilon_a}; \quad \tau_p \gg \tau_t$$

where υ_a is the acoustic velocity in the liquid, it is possible to determine the expression for the detected pressure. The previous limitations indicate that a suitable light source is offered by flash lamp pumped dye lasers where $\tau_p \sim 1\mu sec$. Due to the condition $\alpha\ell \ll 1$ the absorbed energy can be considered to be uniformly distributed along ℓ, so that assuming an adiabatic and isobaric expansion we can consider the illuminated region as a source of cylindrical acoustic waves. Following Landau procedure, Patel and Tam give the explicit expression for the detectable pressure in the form:

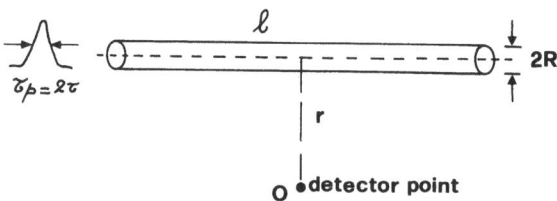

Fig. 4. Geometry for pulsed P. A. technique.

$$p(r,t) = \frac{\rho \upsilon_a}{2\pi} \int_{-\infty}^{t-\frac{r}{\upsilon_a}} \frac{2\pi R \Delta \ddot{R}(t')dt'}{[\upsilon_a^2(t-t')^2-r^2]^{1/2}} \tag{24}$$

where ρ is the liquid specific mass and ΔR is the radius change due
to the expansion. The Eq. (24) gives the causality expression
relating the detectable pressure change and the expansion velocity
rate of the source at the delayed time t'. Assuming for the laser
pulse a gaussian shape with width 2τ and energy E_0

$$\frac{dE(t')}{dt'} = \frac{E_0}{\pi^{1/2}\tau} e^{-(\frac{t'}{\tau})^2}$$

where $E(t')$ is the integrated energy up to time t' and $E(\infty) = E_0$;
if no heat diffusion happens during the laser pulse duration the
optoacoustic equation for $p(r,t)$ can be written as:

$$p(r,t) = - \frac{\upsilon_a \beta \alpha E_0}{\pi^{3/2}c_p\tau^3} \int_{-\infty}^{t-\frac{r}{\upsilon_a}} \frac{t'e^{-(\frac{t'}{\tau})^2}dt'}{[\upsilon_a^2(t-t')^2-r^2]^{1/2}}$$

where β is the volume thermal expansion coefficient and c_p the
specific heat at constant pressure. By evaluating the integral in
Eq. (25) we can observe that the pressure signal, that is detectable
after a delay time $\tau_d = \frac{r}{\upsilon_a}$ from the laser pulse excitation, has a
form of a compressional pulse followed by a rarefaction one separated
by a time that is about τ_p (Figure 5). The above result has been
derived under the assumption $R \ll r$; in the opposite case the illuminated region can be decomposed in elementary cylinders whose acoustic
signals will arrive in phase at the observation point if $R \ll \upsilon_a\tau$,
or with interference effect if $R \gg \upsilon_a\tau$. In the latter case we will
also observe a compression and a rarefaction pulse but separated by
the transit time across the source R/υ_a.

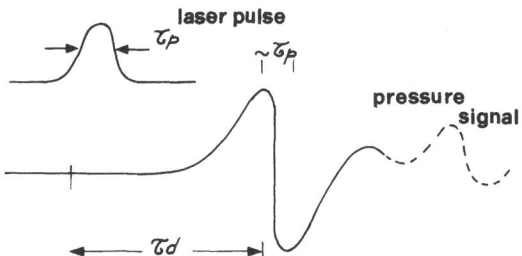

Fig. 5.   Signal shape in pulsed P. A. spectroscopy.

The sensitivity of such optoacoustic technique has been tested by the authors mainly in liquids. They have shown that using laser pulses of ~ 1 mJ/pulse, pulse repetition rate of 10^{-1} sec, integration time 1 sec, and a signal - to - noise ratio of 1, it is possible to measure absorption coefficients as small as $10^{-6} \div 10^{-7}$ cm^{-1}. The main processes that can affect the sensitivity of this technique are: optical absorption from the cell windows, light scattering in the bulk of the sample, electrostriction phenomena in the illuminated region. The signals arising from the first two processes can be minimized by a suitable choice of low loss windows, reduction of scattering impurities, use of highly light reflecting acoustic detectors, and an appropriate time gating of acoustic signals detected by the piezoelectric transducer. As regards the signal from the electrostriction process in the laser irradiated region, for $\tau \sim 1$ μsec it is equal to an acousto-optic pulse obtained with $\alpha \simeq 4.10^{-9}$ cm^{-1}. This means, according to the authors, that the electrostriction effect does not affect the optoacoustic signal value.

The optoacoustic spectroscopy where tunable dye laser source is needed, has been applied to weakly absorbing liquids by using an immersed acoustic detector and gated detection. Linear spectroscopy has been performed for water and transparent organic liquids and, in particular, to measure the absorption profile of high harmonics in C-H stretch of benzene in carbon tetrachloride. In the case of films or powders the sample is placed on a quartz substrate where the acoustic detector is placed at a suitable distance. The substrate form is choiced so to avoid scattered light from the sample reaching to the piezoelectric transducer.

Non-linear spectroscopy has been performed in liquids by means of pulsed optoacoustic technique by Patel and Tam [16]. They have measured two photon absorption in benzene which was of the order of $10^{-52} \div 10^{-51}$ cm^{4}sec molecules^{-1} photon^{-1}. Observations of two photon absorption by optoacoustic methods have been performed for anthracene in alcohol, and POPOP solution by other authors [17].

Pulsed optoacoustic technique also proves to be a relevant tool for Raman-gain spectroscopy [16] due to the fact that the energy deposited in the medium due to the Raman-gain can be measured.

Mordvik and Schlossberg [11] use two transducers in a different arrangement so as to eliminate the effects of light scattering. A similar technique has been employed by Rosencwaig and Willis [18] for P. A. absorption measurements of optical materials and thin films. In practice, although limited in sensitivity by light scattering, they measure absorption in the 10^{-5} range for optical component materials.

APPLICATIONS

Besides P. A. spectroscopy, the main research fields in which
P. A. effect is usefully applied, are listed in the following:

- Surface and bulk absorption coefficient [19,20];
- Non radiative lifetime [21,22];
- Quantum efficiency of fluorescence [23,24];
- Phase transitions [25-27];
- Electron paramagnetic resonance [28-31];
- P. A. imaging and microscopy [32-36];
- Laser damage [37].

REFERENCES

1. A. G. Bell, Am. J. Sci., 20:305 (1880).
2. W. C. Rontgen, Philos. Mag., 11:308 (1881).
3. J. Tyndall, Proc. R. Soc., London, 31:307 (1881).
4. A. Rosencwaig and A. Gersho, J. Appl. Phys., 47:64 (1976).
5. A. Mandelis, Y. C. Teng and B. S. H. Royce, J. Appl. Phys.,
    50:7138 (1979).
6. T. Papa, F. Scudieri and D. Sette, Nuovo Cimento, 1D:129 (1982).
7. R. S. Quimby and W. M. Yen, Appl. Phys. Lett., 35:43 (1979) and
    J. Appl. Phys., 51:1252 (1980).
8. H. C. Chow, J. Appl. Phys., 51:4053 (1980) and J. Appl. Phys.,
    52:3712 (1981).
9. F. A. McDonald, Appl. Phys. Lett., 36:123 (1980) and J. Appl.
    Phys., 52:1462 (1981).
10. R. M. White, J. Appl. Phys., 34:3559 (1963).
11. A. Hordvik and H. Schossberg, Appl. Opt., 16:101 (1977).
12. M. M. Farrow, R. K. Burnham, M. Auzannean, S. L. Olsen, N.
    Purdie and E. M. Eyring, Appl. Opt., 17:1093 (1978).
13. W. Jackson and N. M. Amer, J. Appl. Phys., 51:3343 (1980).
14. A. Hordvik and L. Skolnik, Appl. Opt., 16:2919 (1977).
15. G. H. Brilmyer, A. Fujishima, K. S. V. Santhanam and A. J.
    Bard, Anal. Chem., 49:2057 (1977).
16. An extended review of the technique and results is in C. K. N.
    Patel and A. C. Tam, Rev. Mod. Phys., 53:517 (1981).
17. A. M. Bonch-Bruevich, T. K. Razumova, I. O. Starobogatov, Opt.
    Spektrosk., 42:82 (1977) and Opt. Spektrosk., USSR, 42:45
    (1977).
18. A. Rosencwaig and J. B. Willis, J. Appl. Phys., 51:4361 (1980).
19. H. S. Bennett and R. A. Formann, Appl. Opt., 15:2405 (1976).
20. H. S. Bennett and R. A. Formann, J. Appl. Phys., 48:1432 (1977).
21. L. D. Merkle and R. C. Powell, Chem. Phys. Lett., 46:303 (1977).
22. A. Mandelis and B. S. H. Royce, J. Appl. Phys., 51:610 (1980).
23. R. S. Quimby and W. M. Yen, Opt. Lett., 3:181 (1978).
24. R. S. Quimby and W. M. Yen, J. Appl. Phys., 51:1780 (1980).

25. P. S. Bechthold, M. Campagna and T. Schober, Sol. State Comm.,
    36:225 (1980).
26. M. A. A. Sigueira, C. C. Ghizoni, J. I. Vargas, E. A. Menezes,
    H. Vargas and L. C. M. Miranda, J. Appl. Phys., 51:1403
    (1980).
27. P. Korpium and R. Tilgner, J. Appl. Phys., 51:6115 (1980).
28. R. L. Melcher, Appl. Phys. Lett., 37:895 (1980).
29. R. C. Duvarney, A. K. Garrison and G. Busse, Appl. Phys. Lett.,
    38:675 (1981).
30. H. Coufal, Appl. Phys. Lett., 39:215 (1981).
31. U. Netzelmann, E. v. Goldammer, J. Pelzl and H. Vargas, Appl.
    Opt., 21:32 (1982).
32. Y. H. Wong, R. L. Thomas and G. F. Hawkins, Appl. Phys. Lett.,
    32:538 (1978).
33. H. K. Wickramasinghe, R. C. Ray, V. Jipson, C. F. Quate and
    J. R. Salcedo, Appl. Phys. Lett., 33:923 (1978).
34. Y. H. Wong, R. L. Thomas and J. J. Pouch, Appl. Phys. Lett.,
    35:368 (1979).
35. R. L. Thomas, J. J. Pouch, Y. H. Wong, L. D. Favro, P. K. Kuo
    and A. Rosencwaig, J. Appl. Phys., 51:1152 (1980).
36. G. Busse, Appl. Opt., 21:107 (1982).
37. A. Rosencwaig and J. B. Willis, Appl. Phys. Lett., 36:667
    (1980).

COHERENT ANTI-STOKES RAMAN SCATTERING

J. P. Taran

Office National d'Etudes et de
Recherches Aérospatiales
Châtillon, France

1. INTRODUCTION

The possibility of carrying out temperature and concentration measurements in gases by a Raman spectroscopy analysis was suggested and demonstrated about a decade ago. Following some early publications on this subject [1-2], a massive effort was undertaken in order to evaluate the potential of spontaneous Raman scattering in the important areas of atmospheric sounding and combustion diagnostics. In these experiments, the concentrations of the molecular species are deduced from the intensities of their respective Raman bands and the temperature is obtained from the contour of any one of these bands. Detailed accounts of early experimental work can be found, among other publications, in several Project SQUID and AIAA workshop proceedings [3-5]. Important instrumental developments were accomplished, with improved collection efficiencies and signal to noise ratio enhancement. The fields of applications were rapidly delineated and it appeared that spontaneous Raman scattering could prove valuable for the investigation of such easily analyzed samples as cold or warm aerodynamic jets, but was of limited potential in low pressure gases, fluorescent samples or in luminous reactive media.

A non-linear optical technique capable of performing Raman spectroscopy with much improved signal strength was then proposed as a competitor to spontaneous Raman scattering in the specific area of combustion diagnostics [3-6]. This technique is based on a four wave mixing process called Coherent Anti-Stokes Raman Scattering, or CARS. CARS, which is one of many well-known third-order processes, was actually observed as early as 1963 [7-8], and has been since then applied to crystal spectroscopy [9-11] and to the measurement of third-order susceptibilities in gases [12-14]. Raman spectroscopy by CARS received a considerable impetus in the early seventies when reliable tunable sources of good optical quality were developed. Progress was made in three important areas, which we shall review in turn.

The first area is that of the theoretical understanding of CARS. Particular attention has been paid to the creation of non-linear source polarization, to the birth and growth of the signal electromagnetic wave and to energy exchange processes within the material.

In the second chapter, we discuss the application of CARS to practical temperature and concentration measurements in reactive media.

In the last chapter, we present a prospective study of electronic resonance enhancement in CARS, which shows great promise for a novel form of molecular spectroscopy and for sensitive detection of trace species.

2. CARS THEORY

A. General Presentation

In gases, CARS is observed when two collinear light beams with frequencies ω_1 and ω_2 (hereafter called laser and Stokes respectively, with $\omega_1 > \omega_2$), traverse a sample with a Raman active vibrational mode of frequency $\omega_v = \omega_1 - \omega_2$. A new wave is then generated at the anti-Stokes frequency $\omega_3 = \omega_1 + \omega_v = 2\omega_1 - \omega_2$ in the forward direction, and collinear with the pump beams (Figure 1). This new wave results from the inelastic scattering of the wave at ω_1 by the molecular vibrations, which are coherently driven by the waves at ω_1 and ω_2 (hence the name of the effect). We note that the same mechanism creates, for reasons of symmetry, a similar wave at $2\omega_2 - \omega_1$ (CSRS

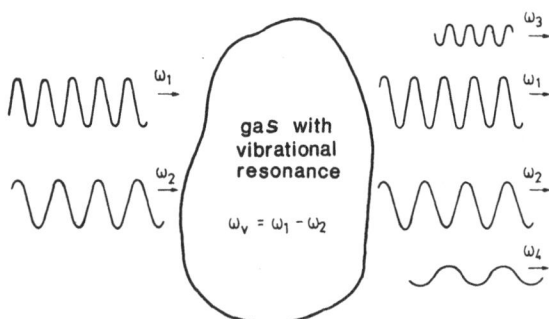

Fig. 1. CARS and CSRS.

for Coherent Stokes Raman Scattering). This wave has been observed and is sometimes used for spectroscopic purposes, in spite of the difficulties connected with background light rejection and poorer detector efficiencies. In the following, we shall give a presentation of the CARS theory and a description of the physical mechanisms.

We recall that, because of non-linearities, it is customary to write the polarization created in a medium by intense optical beams in terms of a power series in the field amplitudes. We have at point $\underline{r}$:

$$\underline{S}(\underline{r},t) = \underline{S}^{(1)}(\underline{r},t) + \underline{S}^{(2)}(\underline{r},t) + \ldots + \underline{S}^{(n)}(\underline{r},t) + \ldots \qquad (1)$$

The main linear contribution $\underline{S}^{(1)}(\underline{r},t)$ and those of higher order (which are smaller in general) all can be given an explicit expansion. To this end, we take the applied radiation in the form of mono-chromatic plane waves. Under steady state conditions, we can expand the total field vector $\underline{\xi}$ as a function of its m distinct frequency components ω_i of wave vector $\underline{k}_i$:

$$\underline{\xi} = \Sigma_i \; \underline{\xi}_i(\underline{r},t) \qquad (2)$$

$$\underline{\xi}_i(\underline{r},t) = \frac{1}{2} \underline{E}(\underline{r},\omega_i)e^{-i(\omega_i t - \underline{k}'_i \underline{r})} + \text{cc.} \qquad (3)$$

The higher-order frequency components of the polarization expansion (1) now can be written. If ω_s is the frequency of one such component, we write the latter as:

$$\underline{S}^{(n)}(\underline{r},t,\omega_s) = \frac{1}{2} \underline{P}^{(n)}(\underline{r},\omega_s)e^{-i(\omega_s t - \underline{k}'_s \underline{r})} + \text{cc,} \qquad (4)$$

with the phenomenological expansion:

$$\underline{P}^{(n)}(\underline{r},\omega_s) = (\tfrac{1}{2})^{(n-1)} \; \underline{\underline{\chi}}^{(n)}(-\omega_s,1_1\omega_{j_1},1_2\omega_{j_2},1_n\omega_{j_n})$$

$$\times \; \underline{E}_{1_1}(\underline{r},j_1)\underline{E}_{1_2}(\underline{r},j_2) \cdots \underline{E}_{1_n}(\underline{r},j_n) \qquad (5)$$

and with $\omega_s = \sum\limits_{i=1}^{n} 1_i\omega_{j_i}$, $\underline{k}'_s = \sum\limits_{i=1}^{n} 1_i\underline{k}_{j_i}$; here, $\underline{\underline{\chi}}^{(n)}$ is the suscepti-bility tensor of order n (the rank of this tensor is n + 1); we also specify $1_i = \pm 1$ and $1 < j_i < m$, and

$$\underline{E}_{1_i}(\underline{r},j_i) = \underline{E}(\underline{r},\omega_i) \; \text{if} \; 1_i = +1$$

$$\underline{E}^+(\underline{r},\omega_i) \; \text{if} \; 1_i = -1;$$

$\underline{S}^{(1)}$ is associated with linear effects (dispersion and absorption;
$\underline{S}^{(2)}$, which is responsible for such effects as frequency doubling or
parametric conversion, vanishes in media possessing inversion symmetry, e.g., centrosymmetric crystals, gases and liquids; all other
even order terms also vanish in these media; $\underline{S}^{(3)}$ stands for a large
class of effects such as third harmonic generation, and three-wave
mixing via two photon and Raman non-linearities.

The source polarization component of frequency ω_s gives birth
to an electromagnetic wave at ω_s. This wave is a solution of the
wave equation, which can be written:

$$(\nabla^2 - \frac{n^2}{c^2}\frac{\partial^2}{\partial t^2}) \underline{\xi}_s(\underline{r},t) = \frac{4\pi}{c^2}\frac{\partial^2}{\partial t^2} \underline{S}^{(n)}(\underline{r},t,\omega_s) \qquad (6)$$

for a non magnetic homogeneous medium. We assume here that $\underline{S}^{(n)}(r,\omega_s)$
as given in Eqs. (4) and (5) is the only source term at frequency
ω_s. Its spectral properties or, in other words, the spectral depen-
dence of the nonlinear optical susceptibility tensor $\underline{\underline{\chi}}^{(n)}(-\omega_s,1_1\omega_{j_1},$
$1_2\omega_{j_2}, \ldots 1_n\omega_{j_n})$ as a function of the applied field frequencies
ω_1, ω_2, etc., are directly reflected in the rate of growth of the
signal wave. By tuning the applied field frequencies and monitoring
the resultant changes in the signal amplitude, one performs a non-
linear optical spectroscopy of the medium. For instance, three-wave
mixing spectroscopy when carried out in the proper frequency domain
gives information about the Raman active vibrational modes of the
medium; this is the basis for a whole line of experiments comprising
pure Raman spectroscopy, analytical chemistry and pressure induced
resonances.

There are numerous instances where the source polarization component of Eq. (6) is not unique. In effect, one can often find other
terms at the same frequency, with the same order in the non-linearity,
but which proceed from a distinct physical mechanism and are associated with a different susceptibility tensor. Terms with a higher
order in the non-linearity are also possible. All such terms giving
contributions at ω_s have to be added to $\underline{S}^{(n)}(\underline{r},\omega_s)$ in the right hand
side of Eq. (6).

For instance, if two fields at ω_1 and ω_2 are applied with $\omega_1 >$
$> \omega_2$, one has two third-order polarization terms at the frequency
$\omega_1 = 2\omega_1 - \omega_2$

$$\underline{S}^{(3)}(\underline{r},t,\omega_3) = \underline{S}^{(3)CARS}(\underline{r},t,\omega_3) + \underline{S}^{(3)SRS}(\underline{r},t,\omega_3). \qquad (7)$$

The first one is the CARS component:

$$\underline{S}^{(3)CARS}(\underline{r},t,\omega_3) = \frac{1}{2} \underline{P}^{(3)CARS}(\underline{r},\omega_3) e^{i(\underline{k}_3'\underline{r}-\omega_3 t)} + cc \qquad (8)$$

with

$$\underline{P}^{(3)CARS}(\underline{r},\omega_3) = \frac{1}{4} \underline{\underline{X}}^{(3)CARS}(-\omega_3,\omega_1,\omega_1-\omega_2) \; \underline{E}^2(\underline{r},\omega_1)$$
$$\underline{E}^+(\underline{r},\omega_2)$$

and $\underline{k}_3' = 2\underline{k}_1 - \underline{k}_2$, while the second one reflects the stimulated Raman scattering (inverse Raman scattering) interaction between the waves at ω_3 and ω_1:

$$\underline{S}^{(3)SRS}(\underline{r},\omega_3) = \frac{1}{4} \underline{\underline{X}}^{(3)SRS}(-\omega_3,\omega_1,-\omega_1,\omega_3)|\underline{E}(\underline{r},\omega_1)|^2 \qquad (9)$$

$$x \; \underline{E}(\underline{r},\omega_3) \; x \; e^{i(\underline{k}_3\cdot\underline{r}-\omega_3 t)} + cc.$$

The latter is negligible in a CARS experiment, since the susceptibility components are of comparable magnitude and since

$$|\underline{E}(\underline{r},\omega_3)| \ll |\underline{E}(\underline{r},\omega_1)|, \; |\underline{E}(\underline{r},\omega_2)|; \qquad (10)$$

k_3 is here the wave vector of the anti-Stokes wave. Similarly, the third-order polarization terms at ω_1 and ω_2 can also be broken down into equations similar to Eq. (6). For these, however, the stimulated Raman scattering term is the stronger.

In conclusion, we have two separate problems to solve in non-linear optical spectroscopy:

- derivation of all the relevant non-linear susceptibility terms;
- calculation of the electric field solution of the wave equation.

The derivation of the non-linear susceptibility terms, which is essential in predicting the spectral properties of a medium, can be done through several distinct appraoches. In the case of third-order Raman-type non-linearities, the classical Placzek model of molecular polarizabilities leads to a rapid calculation of essential results. It gives good insight into the physical mechanisms, but is inadequate for the case where one or more of the light waves is in resonance with one-photon absorption frequencies of the Raman-resonant species. Quantum mechanical derivations are more accurate. Those based on a wave function representation are often sufficient and have been employed extensively using a perturbative treatment of the electric field interactions. We prefer the density operator formalism which, in association with a Feynman-like diagrammatic representation, leads to a rapid derivation of all relevant susceptibility terms and to an easy interpretation of the physical mechanisms involved. The tensor properties of the susceptibility components also follow easily.

The search for the wave equation solution is the second major problem. This solution reveals the important properties of the signal generation: phase-matching, energy exchange between the light waves and the matter, pulse shape characteristics and spatial resolution of CARS measurements using focused beams.

B. Derivation of the Susceptibility

Our purpose in this section is not to give a complete derivation of the susceptibility, but only an outline of the principles. The quantum state of the scattering molecules is represented, at point r, as is conventional, by the density operator ρ with the well-known equation of motion:

$$\frac{\partial}{\partial t} \rho(\underline{r},t) = -\frac{i}{h} [H_0 + V(\underline{r},t), \rho(\underline{r},t)] + \frac{\partial \rho}{\partial t}\Big|_{damp} \qquad (11)$$

H_0 is the free molecule Hamiltonian with a discrete spectrum of eigenstates $|n\rangle$ corresponding to eigenenergies $h\,\omega_n$; the Hamiltonian describing the interaction of the molecules with the radiation field is $V(\underline{r},t) = -\underline{P}\cdot\underline{E}(\underline{r},t)$ in the dipolar approximation; $\underline{P}$ is the dipole moment operator; $\frac{\partial \rho}{\partial t}\Big|_{damp}$ is the damping term, which is determined by stochastic processes such as spontaneous emission of light and collisions between molecules.

We assume the perturbation $V(\underline{r},t)$ to be weak enough to allow the solution of Eq. (11) to be expanded in successive powers of $V(\underline{r},t)$. The density operator is then obtained to any order 1 by the familiar series expansion

$$\rho(\underline{r},t) = \rho^{(0)}(\underline{r},t) + \rho^{(1)}(\underline{r},t) + \ldots + \rho^{(1)}(\underline{r},t) \qquad (12)$$

The 1th order term $\rho^{(1)}(\underline{r},t)$ is proportional to $V^1(\underline{r},t)$ and is obtained by 1 iterative applications of Eq. (11). The term responsible for the CARS polarization is of order 3, and the polarization is given by:

$$\underline{S}^{(3)CARS}(\underline{r},t,\omega_3) = NTr[\rho^{(3)CARS}(\omega_3,\underline{r},t)\underline{P}] \qquad (13)$$

where $\rho^{(3)CARS}(\omega_3,\underline{r},t)$ labels the CARS Fourier component of $\rho^{(3)}(\underline{r},t)$ at frequency ω_3. Identification between Eqs. (8) and (13) eventually yields the expression for the CARS susceptibility tensor.

The entire derivation is straight forward but time consuming. Recently, diagrammatic representations of all possible density operator evolutions have been introduced for the treatment of non-linear optical processes. These representations give useful insight into the microscopic physical mechanisms [15-24]. They are applied with a set of simple rules which allow one to rapidly calculate all the relevant susceptibility terms [17-24].

Similar representations are used in nuclear physics. In our representation, we use the fact that the density operator at any specified order can be shown to result from a number of contributions; each of these is associated with a specific time sequence of perturbations to the density operator, or to the ket vector $|\psi\rangle$ and its complex conjugate $\langle\psi|$ (in the pure state case); the time-ordering of the perturbations to $|\psi\rangle$ with respect to those to $\langle\psi|$ is of crucial importance in the case of collisional relaxation. Each of these elementary time-ordered contributions can be visualized by means of a double-sided Feynman-like diagram. Ordinary Feynman diagrams [25] have been used in non-linear optics [26-29]. However, their application is limited to the case where simplifying assumptions on collisional rate are made [17,18,30] and they do not depict the physical processes as clearly. In a double-sided diagram, the time evolution of the density matrix is depicted along two parallel vertical bars (one for each subscript of the density matrix) with time increasing upwards. Each interaction with the electromagnetic field is represented by a segment pointing downwards from a vertex if it corresponds to a term oscillating as $e^{-i\omega_j t}$ in the interaction Hamiltonian $V(t)$ and pointing upwards if the term oscillates as $e^{+i\omega_j t}$. The vertex is on the left or right hand side vertical bar depending on whether the left or right hand side subscript of the density matrix element is changed through the interaction. The eigenstates between which the interaction Hamiltonian is operating are indicated below and above each vertex.

In CARS, one must combine two vertices at ω_1 and one at ω_2 in order to get the polarization component $\underline{P}^{(3)}$ (as given in Eq. (8)); its $e^{-i\omega_j t}$ dependence implies two segments pointing down from the vertex for the interactions with ω_1 and one pointing up for ω_2. If we use a set of four molecular levels a, b, n, n' as shown in the energy level diagram of Figure 2 (n and n' being of parity opposite to that of a and b), and if the sequence of interactions is applied to be unperturbed density operator $\rho_{aa}^{(0)}$, then there are 24 time-ordered possibilities for this sequence of interactions. Each possible time sequence of interactions gives a distinct contribution to the susceptibility. Another set of 24 terms proportional to $\rho_{bb}^{(0)}$ is also found if state b is populated. It is beyond the scope of this overview to present a detailed account of the rules one uses to derive the susceptibility term from its associated diagram. These rules are found in refs. [17-24] together with their justification.

The CARS susceptibility can be written

$$\underline{\underline{\chi}}^{(3)CARS}(-\omega_3,\omega_1,\omega_1,-\omega_2) = \underline{\underline{\chi}}_{NR} + \underline{\underline{\chi}}_R^{a,b,n,n'} \tag{14}$$

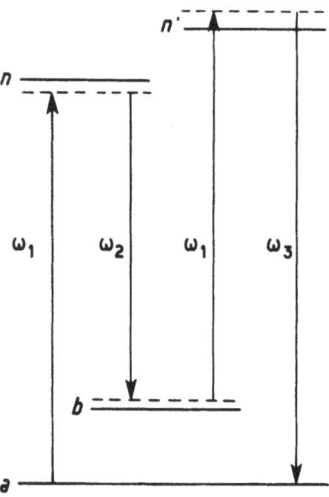

Fig. 2. CARS energy level diagram showing the transitions in
closest resonance with the fields.

where $\underline{\underline{X}}_R^{a,b,n,n'}$ is the Raman resonant part associated with the Raman
active transition of frequency ω_{ba} between a and b, for the set of
two levels n and n'. The tensor element pertaining to a set of
particular field polarizations $\hat{e}_1$, $\hat{e}_2$, $\hat{e}_3$ is:

$$\hat{e}_3 \; \underline{\underline{X}}_R^{a,b,n,n'} \; \hat{e}_1 \hat{e}_2 = \frac{N}{h^3} \; \frac{1}{(\omega_{ba}-\omega_1+\omega_2-i\Gamma_{ba})} \; x \; (A+B)$$

$$x \; [\rho_{aa}^{(0)} (\alpha+\beta) - \rho_{bb}^{(0)} (\gamma+\delta)] \qquad (15)$$

with $A = \mu_{an'}\mu_{n'b}(\omega_{n'a}-\omega_3-i\Gamma_{n'a})^{-1}$,

$\quad\; B = \mu_{an'}\mu_{n'b}(\omega_{n'b}+\omega_3+i\Gamma_{n'b})^{-1}$,

$\quad\; \alpha = \mu_{bn}\mu_{na}(\omega_{na}+\omega_2-i\Gamma_{na})^{-1}$,

$\quad\; \beta = \mu_{bn}\mu_{na}(\omega_{na}-\omega_2-i\Gamma_{na})^{-1}$,

$\quad\; \gamma = \mu_{bn}\mu_{na}(\omega_{nb}-\omega_2+i\Gamma_{nb})^{-1}$,

$\quad\; \delta = \mu_{bn}\mu_{na}(\omega_{nb}+\omega_1+i\Gamma_{nb})^{-1}$.

Here, N is the number density of active molecules. The absorption
frequencies from states $|a>$ and $|b>$ to state $|n>$ are ω_{na} and ω_{nb}
respectively, and the Γ's are the corresponding damping factors;
μ_{an} is the matrix component of the dipole moment operator $\mu_{an} =$
$= <a|\underline{P}\cdot\hat{e}_1|n>$ where $\hat{e}_1$ is the unit vector in the direction of the

polarization of the   field; μ_{bn}, $\mu_{n'b}$, μ_{an}, involve interactions
with ω_2, ω_1, ω_3 fields respectively. If more than four levels are
present, a summation must be taken and the vibrationally resonant
part becomes

$$\underset{=}{X}_R = \underset{a,b,n,n'}{\Sigma} \; \underset{=}{X}_R^{a,b,n,n'}$$

Molecular spectroscopy by CARS consists in carrying out an
analysis of the spectral properties of $\underset{=}{X}^{(3)CARS}(-\omega_3,\omega_1,\omega_1,-\omega_2)$. As
As shown in Eq. (14), the latter contains two parts. In mixtures,
part $\underset{=}{X}_{NR}$, which is called non-resonant, is contributed both by
probed molecules and by the non-Raman-resonant molecular species
(diluent molecules). It is composed of terms analogous to those of
$\underset{=}{X}_R^{a,b,n,n'}$ but with non-resonant two-photon sum or difference denominators in place of the Raman resonance denominator. In the usual
case where the number density N of the probed molecules is small
compared to that of the diluent molecules, $\underset{=}{X}_{NR}$ is mainly contributed
by the latter and is therefore a frequency-independent real tensor
(provided that there are no one- or two-photon electronic resonances
in the diluent molecules). The presence of this non-resonant part
is one of the most severe problems in the application of CARS spectroscopy.

We are particularly interested in the spectral properties of
the Raman-resonant part (Eq. (15)). Off electronic resonance, i.e.,
ω_1, ω_2, $\omega_3 \ll \omega_{na}$, ω_{nb}, $\omega_{n'a}$, $\omega_{n'b}$, all the coefficients A, B, α, β,
γ, δ are of similar magnitude and depend only weakly on the electric
field frequencies. If we assume for simplicity, that all fields
have the same polarization $\hat{e}_1$, the relevant tensor element then
reduces to

$$\hat{e}_1 \; \underset{=}{X}_R^{(3)} \hat{e}_1 \hat{e}_1 \hat{e}_1 = \underset{ab}{\Sigma} \; \frac{Nc^4}{h^3\omega_1\omega_2^3}(\rho_{aa}^{(0)}-\rho_{bb}^{(0)}) \; \frac{d\sigma}{d\Omega} \; \frac{1}{\omega_{ba}-\omega_1+\omega_2-i\Gamma_{ba}} \qquad (16)$$

where $\frac{d\sigma}{d\Omega}$ is the spontaneous Raman scattering cross section. The
spectral analysis thus reveals the Raman resonances ω_{ba} contained in
the denominator of Eq. (16). Identification of these resonances and
monitoring of their amplitude are active research areas for analytical
chemistry. Furthermore, the other tensor elements of $\underset{=}{X}_R$ can be
measured by an adequate choice of field polarizations. All these
properties will be discussed in Chapter 3.

When the electronic resonances are approached, only two terms
in Eq. (15) become large. Thus we have

$$\underset{=}{X}_R \simeq \frac{N}{h} \; \frac{1}{\omega_{ba}-\omega_1+\omega_2-i\Gamma_{na}}(\rho_{aa}^{(0)} A\beta - \rho_{bb}^{(0)} A\gamma). \qquad (17)$$

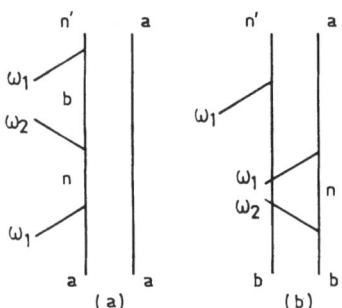

Fig. 3. Time ordered diagrams representing the density operator
         evolutions which lead to the two most important resonant
         CARS susceptibility terms shown in Eq. (16):

         a) term proportional to $\rho_{aa}^{(0)}$;

         b) term proportional to $\rho_{bb}^{(0)}$.

         Indices a, b, n, n' refer to molecular eigenstates as
         depicted in Figure 2.

We give the time-ordered diagrams from which these two terms have
been obtained in Figure 3. Since A, β and γ all undergo large
enhancements and large variations as the field frequencies are varied,
the spectral analysis is complicated somewhat. This particular
problem of resonance-enhanced CARS is treated in Chapter 4.

    In addition to these main vibrationally resonant terms given in
Eq. (15), one has to consider terms which contain vibrational resonances in the excited electronic state and which are generally left
out with $\underline{\underline{X}}_{NR}$. These terms have been given attention recently. They
have been identified and treated as corrections to the main terms in
Eq. (15) by Druet et al. and Yee et al. [17,18,24]. Their spectroscopic and physical importance has been recognized by Bloembergen
and co-workers [30-33] who drew attention to the fact that they
represent vibrational contributions from states that have no initial
population, and who experimentally demonstrated their existence in
Na vapor; yet this existence has been disputed by Eesley [29] and
Carreira et al. [34] who used a simpler quantum mechanical derivation.

    An example of these corrective terms has been treated by Druet
and Taran [24,Appendix I]. This example is different from that of
Bloembergen et al. [33] but also lends itself to experimental checks
in molecular spectroscopy. It is based on the two terms shown in
Figure 4 (with their corresponding resonance denominators) which can
be combined with the one in $\rho_{bb}^{(0)}$ of Figure 3, yielding a susceptibility contribution of the form:

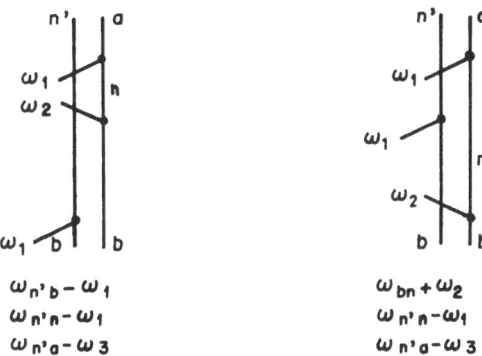

Fig. 4.   Diagrammatic representation of $\rho_{bb}^{(0)}$ contributions to $\underline{\underline{\chi}}^{(3)}$CARS having the resonant denominator $\omega_{n'n} - \omega_1 + \omega_2$.

$$N\rho_{bb}^{(0)} \frac{\mu_{an'}\mu_{n'b}\mu_{bn}\mu_{na}}{h^3} \frac{1}{(\omega_{bn}+\omega_2-i\Gamma_{bn})(\omega_{ba}-\omega_1+\omega_2-i\Gamma_{ba})}$$

$$x \; \frac{1}{\omega_{n'b}-\omega_1-i\Gamma_{n'b}} \qquad (18)$$

$$x \; 1 + \frac{i(\Gamma_{n'a}-\Gamma_{n'b}-\Gamma_{ba}) + i(\Gamma_{nn'}-\Gamma_{n'b}-\Gamma_{bn})\dfrac{\omega_{ba}-\omega_1+\omega_2-i\Gamma_{ba}}{\omega_{n'n}-\omega_1+\omega_2-i\Gamma_{n'n}}}{\omega_{n'a}-\omega_3-i\Gamma_{n'a}}$$

where the terms have been grouped as in Ref. [30].

We can see that the 3 terms in Eq. (18) sum to give one term when there is no damping (= 0) or if:

$$\Gamma_{n'a}-\Gamma_{n'b}-\Gamma_{ba} = \Gamma_{nn'}-\Gamma_{n'b}-\Gamma_{bn} = 0. \qquad (19)$$

This particular equation is what we call the damping approximation. It is satisfied when there is no collisional elastic broadening and if the lifetimes of states $|a>$ and $|b>$ are much longer than those of $|n>$ and $|n'>$. Combining Eq. (18) and Eq. (19), we then obtain:

$$N\rho_{bb}^{(0)} \frac{\mu_{an'}\mu_{n'b}\mu_{bn}\mu_{na}}{h^3}$$

$$x \; \frac{1}{(\omega_{bn}+\omega_2-i\Gamma_{bn})(\omega_{ba}-\omega_1+\omega_2-i\Gamma_{ba})(\omega_{n'b}-\omega_1-i\Gamma_{n'b})} \qquad (20)$$

We notice that the resonances at $\omega_{nn'} - \omega_1 + \omega_2$ vanish and that the terms have combined to merely result in a shift of the electronic resonance in the $A\gamma\rho_{bb}^{(0)}$ term of Eq. (15) from $\omega_{n'a} - \omega_3$ to $\omega_{n'b} - \omega_1$.

The latter expression Eq. (20) is usually obtained from perturbation theory in the absence of damping, which had led Carreira and Eesley to their conclusions [29,34].

The interest for these terms is not purely "zoological". They allow interesting Raman spectroscopic information to be collected about unpopulated states, i.e., without having to populate those states beforehand; they differ in the nature of their broadening by the Doppler effect [23]; they finally pose delicate problems about their true nature and their relation to collisional broadening, as claimed by Grynberg [35]. For all these reasons, they will constitute an active research area in four-wave mixing spectroscopy for the coming years.

C. Wave Propagation

We treat here the problem of gas phase CARS. The amplitude of the signal wave in CARS is obtained from the wave Eq. (6). We take the pump fields in the form of collinear, travelling plane waves as given by Eq. (3). Looking for an anti-Stokes wave also travelling in the forward direction along the z axis, we can reduce the degree of Eq. (6), to obtain a steady state equation:

$$\frac{\partial}{\partial z} \underline{E}_3 = \frac{i\pi\omega_3}{2c} \mid \underline{\underline{\chi}}^{(3)CARS}(-\omega_3,\omega_1,\omega_1,-\omega_2)\underline{E}_1^2\underline{E}_2^+ e^{i\delta kz}$$

$$+ \underline{\underline{\chi}}^{(3)SRS}(-\omega_3,\omega_1,-\omega_1,\omega_3)\underline{E}_1\underline{E}_1^+\underline{E}_3 \mid \tag{21}$$

with $\delta k = k_3^1 - k_3 = 2k_1 - k_2 - k_3$; we have taken the refractive index $n \simeq 1$ and we assume the gas to be homogeneous ($\underline{\underline{\chi}}^{(3)}$ is independent of $\underline{r}$). We have written for simplicity $\underline{E}_1 = \underline{E}(\underline{r},\omega_1)$, $\underline{E}_2 = \underline{E}(\underline{r},\omega_2)$, $\underline{E}_3 = \underline{E}(\underline{r},\omega_3)$ and assumed that $\underline{E}_3$ is a slowly varying function of z:

$$\mid \frac{\partial}{\partial z} \underline{E}_3 \mid \ll k_3 \mid \underline{E}_3 \mid$$

Similar equations also hold at ω_1 and ω_2:

$$\frac{\partial}{\partial z} \underline{E}_1 = \frac{i\pi\omega_1}{2c} [\underline{\underline{\chi}}^{(3)CARS}(-\omega_1,\omega_2,\omega_3,-\omega_3)\underline{E}_1^+\underline{E}_2\underline{E}_3 e^{-i\delta kz}$$

$$+ \underline{\underline{\chi}}^{(3)SRS}(-\omega_1,\omega_3,-\omega_3,\omega_1)\underline{E}_1\underline{E}_3\underline{E}_3^+ \tag{22}$$

$$+ \underline{\underline{\chi}}^{(3)SRS}(-\omega_1,\omega_2,-\omega_2,\omega_1)\underline{E}_1\underline{E}_2\underline{E}_2^+]$$

$$\frac{\partial}{\partial z} \underline{E}_2 = \frac{i\pi\omega_2}{2c} [\underline{\underline{\chi}}^{(3)CARS}(\omega_2,\omega_1,\omega_1,-\omega_3)\underline{E}_1^2\underline{E}_3^+ e^{i\delta kz}$$

$$+ \underline{\underline{\chi}}^{(3)SRS}(\omega_2,\omega_1,-\omega_1,\omega_2) \mid \underline{E}_1^2 \mid \underline{E}_2]. \tag{23}$$

The set of Eqs. (21) - (23) can only be solved numerically. However, if we assume the coupling to be weak, $\underline{E}_1$ and $\underline{E}_2$ can be taken as constants and the SRS term in Eq. (21) can be neglected. Then the latter equation is integrated readily. With boundary condition $\underline{E}_3|_{z=0} = 0$, we have

$$\underline{E}_3 \simeq \frac{i\pi\omega_3}{2c} \underline{\underline{\chi}}^{(3)CARS}(\omega_3,\omega_1,\omega_1,-\omega_2)\underline{E}_1^2\underline{E}_2 e^{i\delta kz} \frac{\sin(\delta kz/2)}{\delta k/2} \qquad (24)$$

$$I_3 = \frac{16\pi^4\omega_3^2}{c^4} |\underline{\underline{\chi}}^{(3)CARS}(-\omega_3,\omega_1,\omega_1,-\omega_2)\hat{e}_1\hat{e}_1\hat{e}_2|^2 I_1^2 I_2$$

$$x \; (\frac{\sin(\delta kz/2)}{\delta k/2})^2 \qquad (25)$$

where $\hat{e}_1$ and $\hat{e}_2$ are the unit polarization vectors of the pump waves. Eqs. (24) and (25) constitute the basis for the interpretation of the anti-Stokes wave properties. The most important ones are the following:

1) the anti-Stokes field polarization vector is oriented along the vector

$$\hat{f}_3 = \underline{\underline{\chi}}^{(3)CARS}(-\omega_3,\omega_1,\omega_1,-\omega_2)\hat{e}_1\hat{e}_1\hat{e}_2,$$

which depends on the applied field polarizations as well as on the tensor properties of the susceptibility. This vector has two independent components associated with the non-resonant and the Raman-resonant parts of the susceptibility (as we have mentioned at the end of Part B). This property can be used for non-resonant background cancellation in the spectra.

2) The CARS signal intensity, which is the parameter directly measured using photodetectors, is proportional to $|\hat{f}_3|^2$ (Eq. (25)), and therefore to the squared number density of the medium. It also has a sinusoidal dependence on z; in gases, however, we have $\delta k \simeq 0$ because the dispersion is weak, so that the behavior is parabolic over long distances. Pump depletion would eventually limit this parabolic growth.

Energy exchange between the light waves can be analyzed by recasting Eqs. (21) - (23) into equations for the rates of change of photon number per unit volume and also considering the rate equation for the molecular population change $N \frac{\partial}{\partial t} (\rho_{aa}^{(4)}-\rho_{bb}^{(4)})$. This discussion has been conducted in detail elsewhere [20-24] and we only summarize the conclusions here for the off-electronic resonance case.

The stronger process is the SRS coupling between the ω_1 and ω_2 waves: one photon at ω_1 is converted into a Stokes photon at ω_2 and a quantum of molecular vibration. The rate of this process is proportional to the imaginary part of $\underline{\underline{\chi}}^{(3)SRS}$. Although it is not specifically a CARS interaction, this process is important because large vibrational population changes can result; this can in turn bring higher order corrections to $\underline{\underline{X}}^{(3)CARS}$ and bias the results.

The CARS generation mechanism per se is made up of two distinct processes:

(i) a "parametric" process whereby two laser photons at ω_1 are converted into a Stokes photon at ω_2 and an anti-Stokes photon at ω_3; the molecules are, on the average, returned to their ground state after the interaction; this process can be reversed if the phases of the waves are changed and its rate is proportional to the real part of the susceptibility;

(ii) a "Raman-like" process whereby a Stokes photon is converted into an anti-Stokes photon and two vibrational quanta, on the average, are taken away from the molecules; this process has a rate proportional to the imaginary part of the susceptibility and can be reversed by changing the phases of the waves.

It is noteworthy that the second process is the only one responsible for the anti-Stokes generation exactly on vibrational resonance, since the real part of the susceptibility then vanishes. Yet, the so-called "parametric" process has often been erroneously cited as being the only CARS mechanism. This belief has originated from the fact that the energy level diagram of Figure 2 gives the misleading impression that the CARS interaction returns the molecules to their initial state. It should be emphasized that such energy level diagrams should be used in non-linear optics to only depict the establishment of polarizations. It is only in the case of processes like Raman scattering or multiphoton absorption that they can also be used to depict energy exchange without ambiguity. Finally the above mentioned considerations on net energy exchange and molecular population changes cannot be dissociated from the quantum processes themselves. In effect, on the microscopic scale, molecules can undergo sequences of interactions which either return them to their initial state after the final interaction with the anti-Stokes field (e.g., Figure 3a) or place them in a different vibrational state (e.g., Figure 3b).

3. PRACTICAL APPLICATION OF CARS

A. General Considerations

The laws governing the signal growth and the spectral properties of CARS have been established in the preceding chapter. We show here how CARS can be used for practical measurements and what level of performance can be obtained.

Spatial resolution. Unfocussed parallel beams with large diameters are seldom used because no spatial resolution is possible in this geometry. Since the growth of the power density I_3 is proportional to $I_1^2 I_2$, it seems advantageous to focus the beams to a small diameter and to use high peak power sources. If the condition $\delta k = 0$ is assumed, then it can be shown that:

1) the anti-Stokes flux is contained within the same cone angle as the pump beams energing from the focal region;

2) this flux is generated for the most part within the focal region (where $I_1^2 I_2$ is large);

3) the total power in the anti-Stokes beam some distance beyond the focus is independent of beam diameter and focal length and is approximately given by

$$P_3 = (\frac{2}{\lambda})^2 \ (\frac{4\pi^2\omega_3}{c^2})^2 |\hat{f}_3|^2 P_1^2 P_2, \qquad (26)$$

where refractive indices were taken as unity, where $\lambda = 2\pi c/\omega$ with $\omega \simeq \omega_1 \simeq \omega_2 \simeq \omega_3$ and where P_1 and P_2 are the powers at ω_1 and ω_2 respectively. This expression was obtained by assuming that all the signal is generated from a small cylindrical volume about the focus having a length equal to the confocal parameter l of the beams. If Gaussian beams are used, the beam waist at the focus is $\phi = 4\lambda f/\pi d$ where f is the focal length of the lens and d the beam diameter in the plane of the lens: we also have $1 = \pi\phi^2/2\lambda$. In reality, 75% of P_3 are generated from a volume of length 6l as shown by numerical calculations.

In practical experiments, the spatial resolution is on the order of 1 to 20 mm with laser beams of good optical quality. This may still be insufficient in some experimental situations where higher spatial resolutions are needed. A particular beam arrangement called BOXCARS has been proposed for better resolution [36]. In this arrangement, the beams are crossed at a small angle so that the polarization wave vector k_3' remains equal to the anti-Stokes signal wave vector k_3 (Figure 5). The beam configuration is shown in Figure 6. The spatial resolution is well under 2 mm.

Spectral information. CARS spectroscopy can be accomplished in various manners depending on the application envisioned (e.g., high-resolution spectroscopy or chemical analysis). The spectra are usually retrieved by holding ω_1 fixed, varying ω_2 so that $\omega_1 - \omega_2$ is swept across the resonances of interest while monitoring the anti-Stokes flux. In gas mixtures, the following information is obtained from the spectra using Eq. (16):

- composition since each molecular species has a particular set of vibrational resonances which can seldom be confused with that of other species;

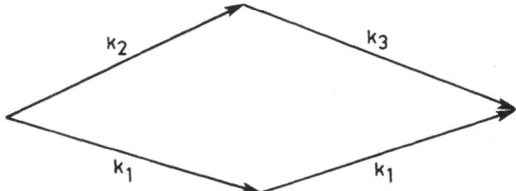

Fig. 5. Wave vector diagram for crossed-beams phase-matched CARS
         or BOXCARS.

     - temperature since the frequency of any particular vibrational
mode depends slightly on the rotational quantum number (in Q-branch
transitions) and on the vibrational quantum number; the resulting
splitting can be used to monitor populations in distinct rovibrational
states and deduce the rotational and vibrational temperatures using
the corresponding Boltzmann coefficients.

     We note, however, that the existence of the non-resonant sus-
ceptibility poses a problem with the detection of trace species in
mixtures, since the non-resonant contribution from the diluent gases
may swamp the Raman-resonant part of the trace species of interest.
As a matter of fact, detection sensitivities are in the range of
10^2 to 10^4 ppm for most cases of interest. These figures can be
improved by a factor of about 30 if advantage is taken of the
different tensor properties of $\underline{\underline{X}}_{NR}$ and $\underline{\underline{X}}_R$ (polarization CARS [37]).

     <u>Advantage of CARS</u>. CARS offers many advantages over other
optical methods for nonintrusive spatially resolved diagnostics of
gases and reactive media:

     - spatial resolution is excellent;

     - the signal is emitted in a well collimated beam, which makes
discrimination against stray light easier;

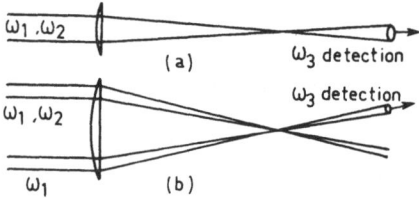

Fig. 6. Experimental beam arrangement: a) conventional CARS;
         b) BOXCARS. In BOXCARS, a conventional CARS beam is also
         emitted in the direction of ω_2; this beam is actually 10
         to 50 times stronger than the BOXCARS beam.

- the spectral position of the signal, to the anti-Stokes side
of the pump, makes it easier to reject fluorescence interference
(which usually lies to the Stokes side of the pump);

- the signal strength is considerable; using conventional pulsed
solid state and dye lasers, the number of photons collected in a
typical experiment is about 10^5 to 10^{15}, i.e., 5 to 10 orders of
magnitude larger than that collected in a spontaneous Raman scattering
experiment.

All these advantages justify the introduction of CARS as a
diagnostics tool in reactive media. This application is to date the
most important one. Other applications, such as high resolution
molecular spectroscopy or chemical analysis of biological samples,
also are attractive but shall not be discussed here.

B. CARS Instrumentation

We describe here the CARS spectrometer in use at ONERA. This
spectrometer was developed jointly with Quantel. The optical components for the laser sources and the beam combining optics are
bolted directly onto a light-weight, portable, rigid 50 cm x 150 cm
cast aluminium table (Figure 7). The passively Q-switched YAG
oscillator with two amplifiers and one frequency doubler delivers
over 150 mJ of radiation at 532 nm in 10 ns pulses at 1 to 10 Hz
(ω_1 beam). The output is single frequency over 95% of the shots and
presents a spectral jitter under ± 0.01 cm^{-1}. These characteristics
are possible only through the use of a stable cavity design for the
YAG oscillator. A second doubler is used to convert the remaining
infra red energing from the first doubler, thus producing an
additional 40 mJ of green to pump the dye chain. This one is composed of a dye laser and one amplifier stage and produces the
"Stokes" beam at ω_2. The dye laser can be tuned with a fixed, high
incidence grating and a rotating mirror. The line-width is 0.7 cm^{-1};
it can be reduced to 0.07 cm^{-1} through insertion of a prism beam
expander. This operation maintains the cavity alignment precisely
and causes a slight shift of the line center.

With a 40 mJ pump energy, the dye chain delivers from 1 to 5 mJ
of tunable radiation in a diffraction-limited beam over the useful
CARS range of 560 - 700 nm, regardless of the line-width. The tuning
is driven by a stepping motor which allows both a continuous sweep
from 500 to 800 nm in coarse steps of 0.07 cm^{-1}, and limited sweeps
of 6 nm about the coarse drive setting in fine steps of 0.007 cm^{-1}.
A broadband mode of operation is also provided for the dye laser,
giving about 100 cm^{-1} line-width; this mode is used for multiplex
CARS experiments in conjunction with a spectrograph and an optical
multichannel analyser [38]. In this case an interference filter is
used for the tuning.

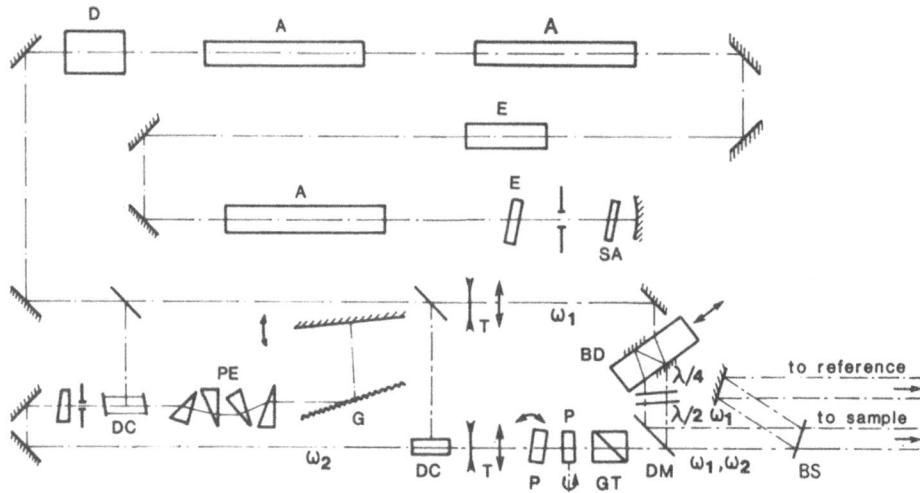

Fig. 7. Laser source assembly. A: Nd YAG amplifier; BD: parallel
         plate for production of parallel beams for BOXCARS (BOXCARS
         arrangement shown, translation of the plate allows passage
         to collinear arrangement without loss of alignment); BS:
         beam splitter for reference channel; D: KDP doubler; DC: dye
         cell; DM: dichroic mirror; E: Fabry-Perot etalon; G: grating;
         GT: Glan-Thompson prism; P: AR coated parallel plate for
         beam translation; PE: prism expander; SA: saturable absorber;
         T: telescope; $\lambda/4$ and $\lambda/2$ are quarter-wave and half-wave
         plates respectively; ω_1: "laser" beam; ω_2: "Stokes" beam.

     A space of 60 cm x 30 cm is left on the table for the mounting
of various beam handling optics. This space is occupied by components for beam-matching and superposition, and for simultaneous
non-resonant background cancellation and improved spatial resolution
using BOXCARS [39]. We thus use telescopes to adapt the divergences
of the beams, parallel plates to translate the beams and to split
them for BOXCARS, one dichroic mirror for beam superposition and
alignment, as well as polarization rotators and polarizers. The
dichroic mirror is held on a sturdy mount having better than 10 μrad
alignment sensitivity. A small fraction (≈5%) of the laser beams
is subsequently split off to pump the reference channel. In the
following, we shall concentrate on the detection system and on the
problem of background cancellation.

     The detection assembly, including the reference leg, is installed
on a separate table of 50 cm x 150 cm (Figure 8). All focusing
lenses are AR coated air-spaced achromats. The anti-Stokes signals
are filtered by means of compact double monochromators preceded by
dichroic filters to prevent breakdown on the entrance and intermediate diaphragms; detection is done with PM tubes which are mounted

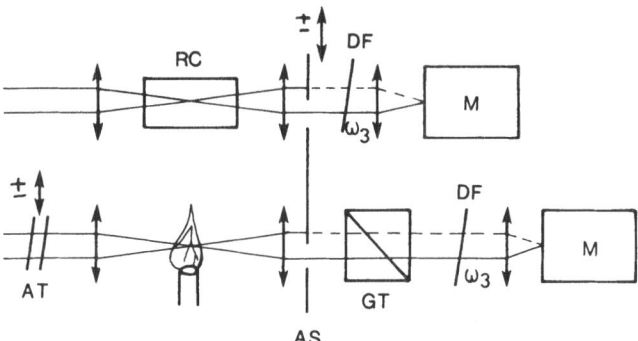

Fig. 8.   Schematic of sample and reference channels. AS: movable
          aperture stop for operation with parallel beams or crossed
          beams (cross-beam position shown here); AT: movable attenu-
          ators; DP: dichroic filters; M: monochromator and detector;
          RC: reference cell.

in the same rack as the signal processing electronics (see below)
to avoid RF interference. Light is piped to them by means of 1 mm
diameter optical fibers. The anti-Stokes signal levels in the
sample and reference channels are adjusted at approximately 10^4
photoelectrons per shot, which corresponds to a Poisson uncertainty
of about 1%. Higher fluxes may cause saturation, lower fluxes
result in unacceptable uncertainty levels. The sample and reference
channels are matched carefully, especially in BOXCARS experiments.
The reference cell contains 50 b of argon. The reference signal is
usually much stronger than needed and has to be attenuated; it
remains adequate even when a combination of BOXCARS and background
cancellation is used (see below).

     In both channels, the signal level is maintained at its pre-
scribed level of 10^4 photoelectrons by adjusting the pump powers
with attenuators, which is a safeguard against Raman saturation at
line center. The photocurrent pulses are treated by an electronic
device that gates them, calculates their ratios, square roots and
average for a fixed number n of shots (n = 1 to 10 in practice).
The electronics unit also rejects shots which do not fall with ±35%
of the mean in the reference leg, and tunes the dye laser after the
n shots have been collected.

     For multiplex CARS, a spectrograph and an optical multichannel
analyzer (OMA2 from PAR) are used. The dispersive element in the
spectrograph is a 2100 lines/mmn, aberration-corrected concave
holographic grating with f = 750 mm. The net spectral resolution
is 0.7 cm^{-1}. Both signal and reference spectra are recorded simultaneously on the vidicon and ratioed channel by channel; square
roots, and averages if necessary, are subsequently calculated.

Recording the reference spectrum is a vital requirement since the
dye laser spectrum is not reproducible and exhibits appreciable
modulation.

The technique of background suppression using the tensor
properties of the non-linear susceptibility has been studied in
detail for collinear beams, and several possible polarization
arrangements have been described [37,40]. In BOXCARS, some flexi-
bility is afforded by the availability of two spatially distinct
"laser" beams at ω_1, which can have different polarizations. A
discussion of that problem will be found in reference [39].

C. Results

The feasibility of concentration measurements in flames by CARS
was shown in 1973 [41]. Since then, many experiments have been
carried out and many species have been detected. Here we shall
concentrate on results obtained at ONERA in recent times.

1. Measurement of rovibrational excitation of H_2 in a discharge.
The study of rovibrational populations in tenuous discharges is an
impossible task using mechanical probes. This study is also exceedingly difficult using absorption/emission methods. CARS, however,
offers an interesting measurement potential over an appreciable
temperature and density range. This was demonstrated recently [42]
in an H_2 discharge (Figure 9) designed for H^- production [43]. The
spectra are shown in Figures 10 and 11 for a total pressure of 0.13
mbar, without the discharge (Figure 10a) and with a discharge voltage
of 90 V and current of 3 A (Figure 10b). Collinear CARS without
background cancellation was used for maximal signal strength and
sensitivity. Note that, since the spatial resolution of collinear
CARS is not excellent, the amplitude of the v = 0 lines may come out
slightly stronger because of some signal contribution from the cold
H_2 surrounding the generator. Only the central portions of the
first four Q lines were plotted. Horizontal bars on the plots give
the theoretical heights of these lines assuming Boltzmann equilibria
at 290°K (Figure 10a) and 475°K (Figure 10b) for the rotation.
Uncertainties in temperature measurements are 5°K and 15°K respectively. From the line intensities with the discharge turned on, we
deduce that molecular H_2 constitutes approximately 90% of the gaseous
mixture, the rest being composed of ions, radicals and electronically
excited H_2. The comparison of the relative amplitudes of Q(1) lines
in v = 0, v = 1 and v = 2 (Figure 11) also gives a measure of the
vibrational excitation.

The line associated with v = 3 could not be detected. In this
preliminary work, the detection sensitivity on v = 2 is about 10^{12}
cm^{-3}. A sensitivity of 10^{11} cm^{-3} is technically feasible and should
be demonstrated in the near future.

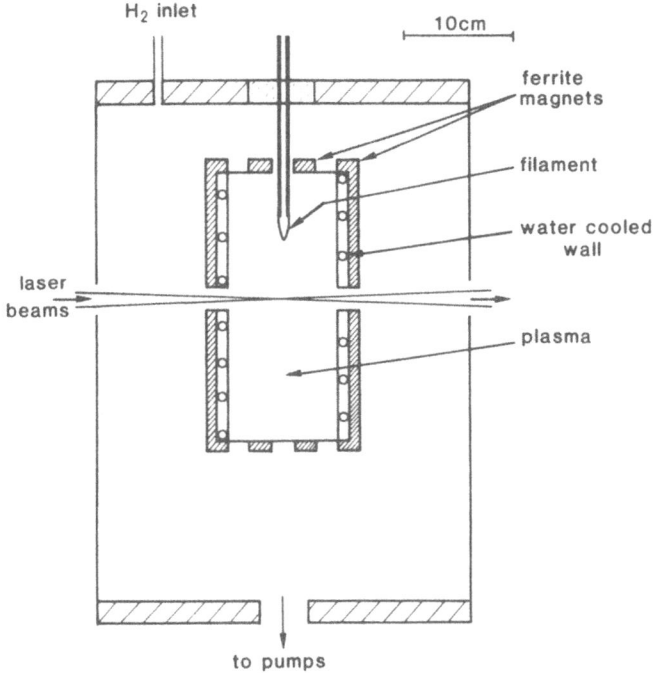

Fig. 9.   Schematic of the plasma generator, drawn to scale.

The present results give:

$$N(v=0) = 1.9 \ 10^{15} \ cm^{-3}$$
$$N(v=1) = 4.5 \ 10^{13} \ cm^{-3}$$

and

$$N(v=2) = 4.4 \ 10^{12} \ cm^{-3}$$

assuming

$$N(v=3) = 10^{12} \ cm^{-3}.$$

The assumption on $N(v=3)$ is necessary, since it is the difference between the number densities of the Raman levels which is measured in CARS experiments (Eq. (15)). These results clearly show that the vibrational equilibrium established is non-Boltzmann (Figure 12). Similar effects were also seen in D_2 and N_2 using CARS [44].

    2.   Temperature and H_2 concentration measurements in an ethylene-air Bunsen flame. H_2 gas plays an important role in the chemistry and energy balance of flames. This was demonstrated recently in a premixed Bunsen flame burning ethylene in air [45]. By recording CARS spectra of the Q-branch of the H_2 formed by the pyrolysis of ethylene, we were able to obtain the static temperature profile and the H_2 concentration distribution. A BOXCARS optical arrangement was used to give a spatial resolution of about 1 mm.

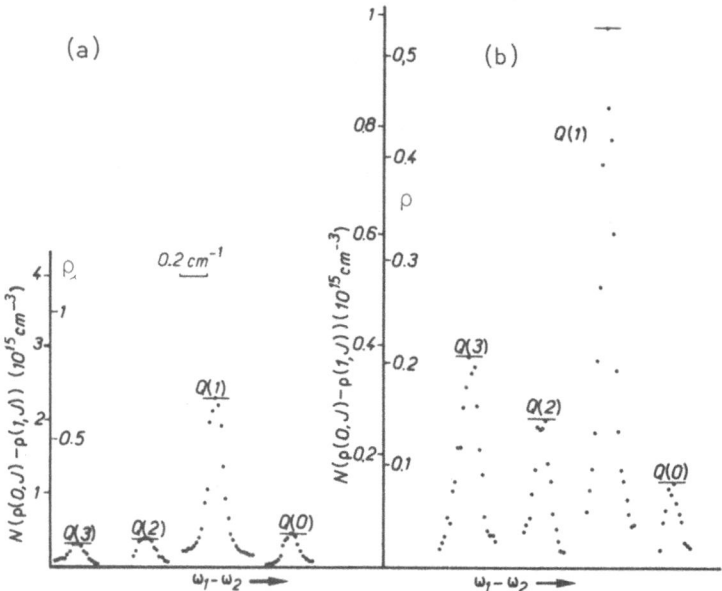

Fig. 10. Profiles of lines Q(0) to Q(4) for the v = 0 → 1 fundamental transition of neutral H_2:
a) without the discharge;
b) with a discharge of 90 V, 3 A.
The strongest portions of the lines only are shown, in
steps of 0.02 cm^{-1}, and with 10 consecutive measurements
averaged at each point. The spectral resolution of the
dye laser was 0.07 cm^{-1}.

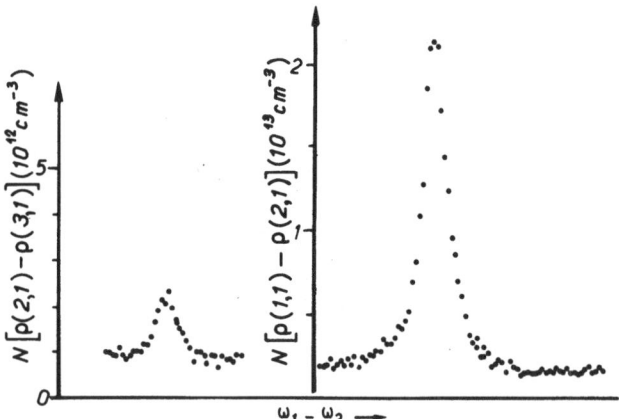

Fig. 11. Profiles of excited states lines, with 30 measurements
averaged at each point.

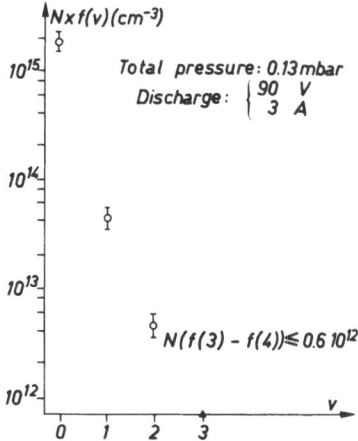

Fig. 12. Distribution of vibrational populations in the discharge.

The spectral analysis was similar to that done in the plasma, although we only monitored peak line intensities. In the data processing, special precautions had to be taken with regard to line broadening mechanisms and concentration calibration. The burner itself was designed so as to give good flame stability. It has a diameter of 10 mm. It was operated at a C/O ratio of 0.57 and mass flow rates of 8.9 mg/sec for C_2H_4 and 76.2 mg/sec for air; this gave a flame cone height of 26 mm at 296°K and 0.98 bar, corresponding to a burning velocity close to 17 cm/sec. The results are presented in Figures 13 and 14.

Most interesting is a comparison of these results with thermodynamic equilibrium calculations of adiabatic flame temperature and stable product composition. Using the water-gas equilibrium and neglecting dissociation, we find for our flame an adiabatic flame temperature of 1940°K and a composition of 14.9 mol% CO, 4.0 mol% CO_2, 10.2 mol% H_2O and 8.6 mol% H_2. There is strikingly good agreement between the final measured temperature in the burnt gas and the calculated temperature, and, in particular, between the hydrogen molar fraction measured just beyond the reaction zone and its thermodynamic equilibrium value of 8.6%. This agreement leaves no doubt that the CARS technique is indeed suitable for spatially resolved concentration measurements in reactive media. However, more checks have to be carried out, e.g., CARS measurements of CO, CO_2 and H_2O to find out if water-gas equilibrium is established or not. Furthermore, the hydrogen rotational temperatures should be compared to N_2 rotational temperatures. Hence, at the present moment, we regard our measured hydrogen concentration as semi-quantitative; it will be improved upon once better knowledge of lineshapes and widths becomes available.

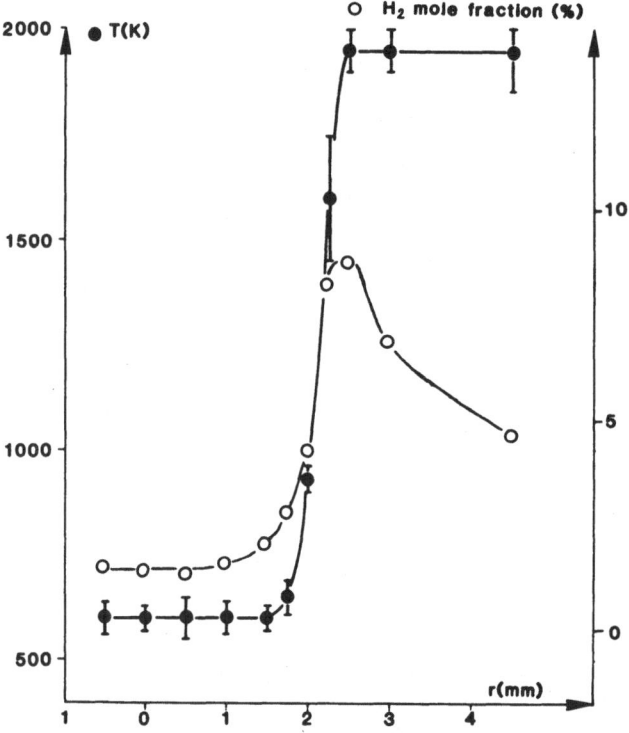

Fig. 13.   Rotational temperatures and hydrogen molar percentages O
           as a function of radial distance from burner axis. Height
           above burner exit plane: 11.5 mm.

     As to now, our results demonstrate remarkably well, besides
oxidation of hydrogen in the burnt gas, the effect of hydrogen
diffusion from the reaction zone into the fresh gas. It is quite
surprising that the hydrogen molar fraction is about 1% at h = 11.5
mm (i.e., 14.5 mm downwards from the flame top and at a radial
distance of 2.5 mm from the flame cone) and as high as 3.7% at a
distance of 2 mm below the flame top. Hydrogen is thus enriched in
the central flow line. A similar effect, the enrichment of heavier
hydrocarbons in the central flow line of a Bunsen flame, was described by Flossdorf et al. [46]. However, the enrichment of
hydrogen in the fresh gas has some implications on the heat balance
in the fresh gas. It is quite conceivable that the pre-reaction
zone hydrogen originates from the recombination of hydrogen atoms.
This would, as a quick calculation shows, for a hydrogen molar
fraction of 1% give a temperature rise of the fresh gas of about
$80°K$ and hence partially account for our observed temperature rise
of about $200°K$, compared to ambient, at h = 11.5 mm. Even if the
molecular hydrogen itself was the main diffusing species, it would
still carry some enthalpy and thus still act as a "heat recirculating

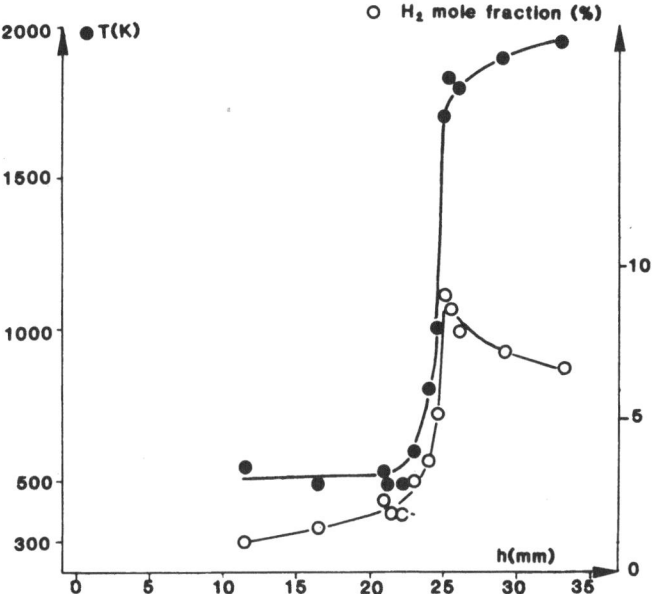

Fig. 14. Rotational temperatures and hydrogen molar percentages 0
          as a function of distance from burner exit plane along
          burner axis.

chemical carrier", though, of course, in a much less efficient way
than the hydrogen atom. The rise in H_2 concentration near the
reaction zone with increasing height above the burner exit plane
could also be an explanation for the well established fact that the
burning velocity is not a constant over the total flame area [46].

    3. Instantaneous measurement of temperature in a well-stirred
reactor. The well-stirred reactor presents interesting challenges
to the CARS diagnostician: can one measure the magnitude of the
temperature fluctuations and the gas residence time? None of these
measurements can be undertaken properly using conventional probing
techniques, yet they are of extreme interest to the theorist.

    A well-stirred reactor was built for these studies [47]. The
reactor is cylindrical in shape with 5.5 cm id. and 3 cm height.
A mixture of methane and air is injected through an array of small
holes placed on a central injection column. Exhaust is through
five 1 cm diameter holes drilled in the top flange.

    The CARS beams are passed horizontally through holes of 6 mm
diameter which are diametrically opposed in the cylinder walls.
The temperature study was done in the multiplex mode, using a BOXCARS
configuration which gave under 1 mm spatial resolution. Numerous

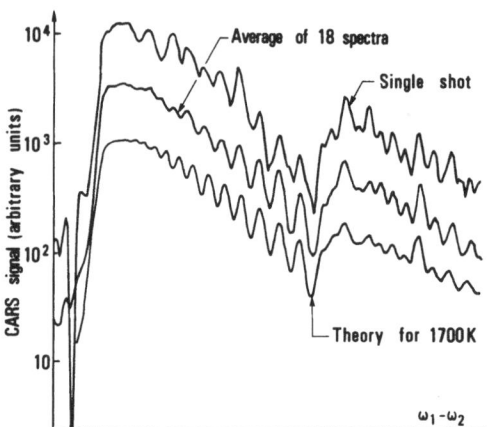

Fig. 15.   Single shot BOXCARS spectrum of N_2 in well-stirred reactor
and calculated spectrum assuming $T = 1700°K$. The peak
amplitude corresponds to about 5.10^3 photoelectrons per
OMA channel. Spectral resolution is about 1 cm^{-1}. Gated
intensifier stage.

instantaneous spectra were recorded. That presented in Figure 15
is typical. This spectrum is shown with a theoretical spectrum
calculated for a temperature of 1700°K. The single shot temperature
measured is close to the adiabatic temperature (2100°K), the drop
being easily accounted for by losses to the walls. Thermocouples
and IR pyrometry both give temperatures close to 1800°K. The
standard deviation in the measured peak intensity (16%) receives
contributions from the local temperature fluctuations as well as
from our measurement uncertainty. The measurement uncertainty comes
largely from beam wandering caused by the gradients in the flame.
Originally, a large source of imprecision was the mixing of hot
flame gases with cold air in the vicinity of the optical ports.
This caused amplitude fluctuations in the range of 1 to 4, and dis-
tortions of the spectral contour which seriously affected the temperature measurement accuracy. The cure was to apply fused silica
windows to the ports. Heating of the walls limits burner operation
to 5 minutes, which enables one to collect about 500 spectra.

The residence time was monitored using a tracer injected in the
fresh gases immediately before the burner for periods of 30 msec
separated by 30 msec. The tracer which we intended to use originally
was CF_4, but this gas was rapidly abandoned because of its chemical
reactivity in the burner above 1500°K, and also because of the rapid
drop in CARS line intensity as a function of temperature due to
broadening (Figure 16). Instead, we used CO_2 which offers two sharp
and intense Q branches which do not broaden as a function of temperature (Figure 17). The dye laser was operated narrowband and

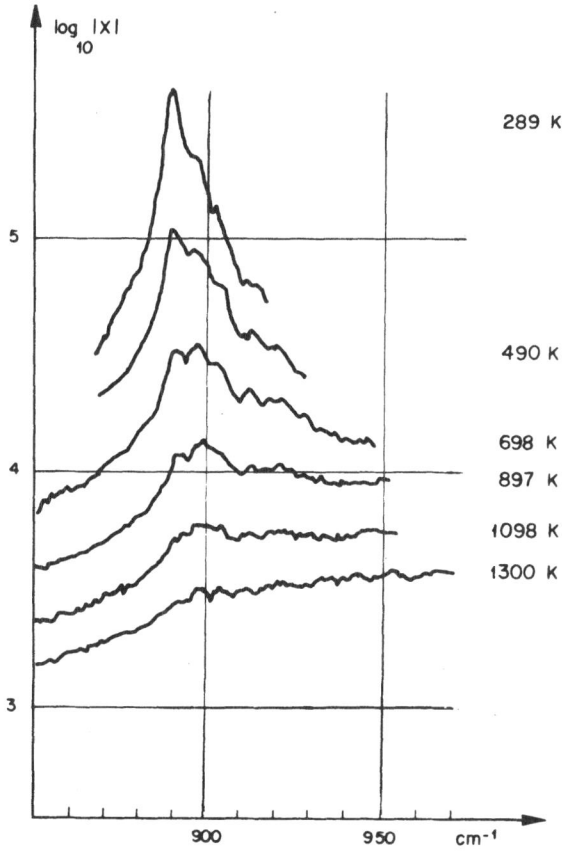

Fig. 16. Spectrum of pure CF_4 in furnace at atmospheric pressure and for several temperatures, recorded at 0.1 cm^{-1} resolution. Scanning CARS was used, with 4 measurements averaged at each point.

was tuned approximately 0.2 cm^{-1} away from the ν_1 mode so that a small frequency jitter would not cause excessive signal change. For a sample of fixed concentration, the standard deviation on the CO_2 concentration measurement is thus 15%. Figure 18 shows the exponential rise of the excess CO_2 concentration in the burner following beginning of its injection. In the well-stirred reactor, the rise and decay times are in principle equal. The residence time deduced from this curve (1.8 msec) is in good agreement with that (1.7 msec) obtained by dividing the burner volume (71 cm^3) by the volumetric flow rate used in that particular run (41 10^3 cm^3/sec at 1700°K using 7.82 g/sec of air and 0.404 g/sec of CH_4); 0.627 g/sec of CO_2 was introduced during the injection periods.

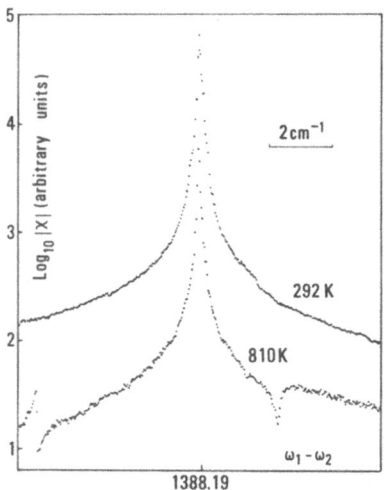

Fig. 17.   Spectrum of pure CO_2 at atmospheric pressure for two
values of temperature.   Same conditions as for Figure 16.

   4.   Temperature measurements in a simulated turbomachine combustor.   Large scale combustors are difficult to tackle by CARS,
since environment is harsh and optical access is limited.   Therefore,
many precautions have to be taken to protect the optical set-up from
noise, dust and oil vapors.   The feasibility of making temperature

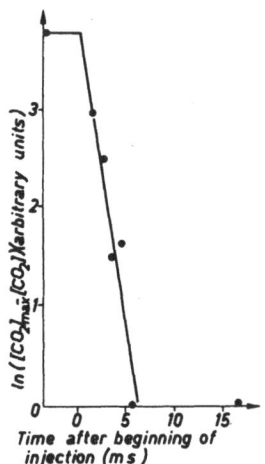

Fig. 18.   Rise in CO_2 concentration in reactor following beginning
of injection; 100 measurements are averaged for each data
point; repetition rate: 2 Hz; the data are presented after
subxtration of maximal CO_2 concentration during injection;
opening and closing times are about 0.3 msec.

and concentration measurements by CARS in a kerozone fueled burner
was tested at SNECMA [48]. The burner produces a flow which is
approximately rectangular, with a size of 15 x 50 cm, and a mass
flow rate approaching 2 kg/sec for atmospheric pressure operation.
The CARS optical tables were installed in a test room near to that
housing the combustor (Figure 19), along with the power supplies
and electronics. Single shot BOXCARS spectra were taken at various
locations in the burner, with a quality approaching, but not as good
as, that of Figure 15. A small percentage of spectra was lost in
the primary zone (near the injectors) because of fuel droplets and
turbulence, which cause spectral envelope distortion and attenuation.

Figure 20 presents a temperature profile recorded in the primary
zone. This temperature is a time average over about 20 consecutive
single shot measurements recorded in a 20 minute period (the data
acquisition rate then was close to 2 per minute). The profile

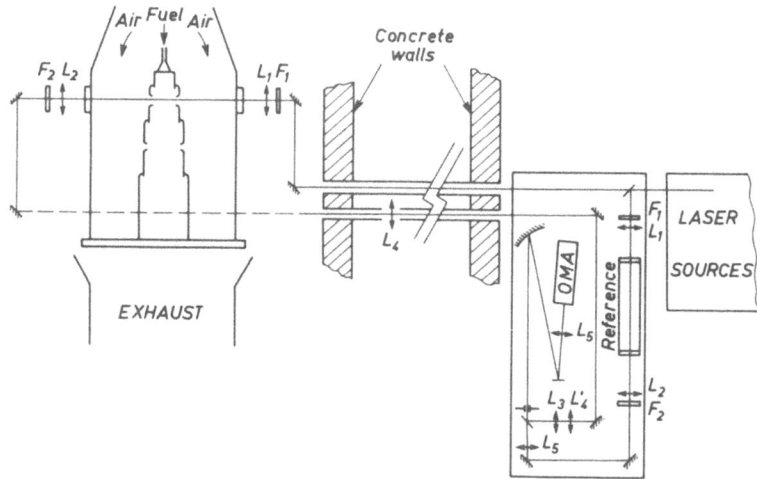

Fig. 19. Layout of optical set up near kerozene fueled combustor.
          The details of the laser source assembly (Figure 7) are
          not given here. The optical path to the burner is about
          10 m. The beams travel several meters through open air
          and had to be enclosed in metal pipes for safety. F_1:
          Schott OG 515 filter; F_2: short-pass dichroic filter for
          rejection of pump beams; L_1, L_2: focusing and recollimating
          achromats; L_3: achromat used to focus anti-Stokes beams
          into entrance slit of spectrograph; L_4, L'_4: a focal lens
          combination used to reduce the effect of beam steering in
          the burner; L_5 gives a magnification of 2 on the OMA face.
          The anti-Stokes beam is passed under the combustor rig
          (dashed line). Access to the primary reaction zones is
          through air dilution holes in the can.

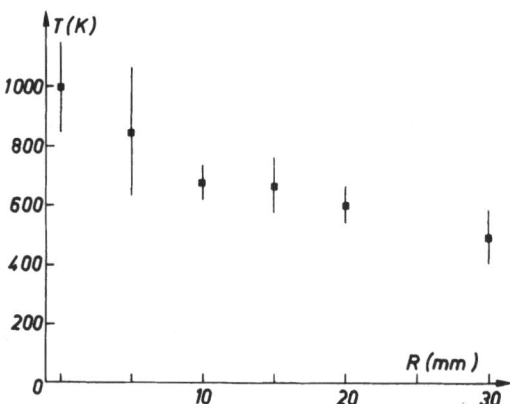

Fig. 20.   Radial temperature profile in primary zone, R is measured
           from burner midplane; note that the pump beams cross the
           plane of the air injection port at R = 50 mm. The error
           bars represent the standard deviation in the measurements
           and reflect contributions from real temperature fluctu-
           ations and from measurement uncertainty.

clearly shows the drop in temperature, which tends to a limiting
value of 500°K close to that of the preheated air used in that
particular run.

We have also studied O_2 and CO, trying to obtain their spatial
distributions. The detectivity using BOXCARS without background
cancellation was about 3% in mole fraction for these species in the
flame, at the time of these experiments. With background cancellation, the detectivity was further degraded by a factor of two.
These results will be published elsewhere [48].

5. Conclusion. Only a few of the potential applications of
CARS have been presented here. Many more have been explored and
demonstrated in other laboratories. Piston engines, jet engines,
chemical vapor deposition (CVD), laser media, all have been or are
currently being probed by conventional CARS. See the contribution
by Stenhouse in this series for a review.

The situation is far different in resonance-enhanced CARS
which, in spite of its extremely promising characteristics, remains
an extremely difficult technique to apply. The next chapter is
devoted to some of its aspects.

4. RESONANCE-ENHANCED CARS

Conventional CARS offers a detection sensitivity of about 1000 ppm in usual gas mixtures near STP. Improvements up to 30 are possible using cancellation of the non-resonant background (polarization CARS) as we have seen in the previous chapter. Much larger improvements on the order of 10^2 to 10^4 are actually obtained using resonant CARS, as we showed with I_2 vapor [20]. This has been the major motivation for several other research programs in resonant CARS throughout the world. Observations in C_2 [49-50] and NO_2 [51-52] have been reported recently. A related area of research, resonance CARS in absorption continua, is treated as a separate topic by Kiefer in this course. This form of spectroscopy has been performed recently in Iodine [53]; such studies are delicate since the susceptibility is much weaker than when enhancement is obtained with discrete lines. We here restrict our discussion to the latter case.

1. Spectral Properties

Resonance CARS spectroscopy is usually carried out by holding ω_1 fixed near a one-photon absorption and by varying ω_2. There are three types of lines associated with the lower state $|a\rangle$ (term proportional to $\rho_{aa}^{(0)}$), as depicted in Figure 21. We call these laser-enhanced Raman resonances (Figure 21a), double-electronic resonances (Figure 21b) and anti-Stokes-enhanced Raman resonances (Figure 21c). Both fundamental and overtone vibrational transitions are possible. In addition, one may find similar sets of resonant transitions associated with $|b\rangle$ if that state is significantly populated; their spectral properties are easily derived from the corresponding resonance denominators.

A notable character of the double-electronic resonances is that their positions in the spectrum, which are given by the condition $\omega_1 - \omega_2 = \omega_{n'a} - \omega_1$, are a function of ω_1 (contrary to Raman resonances of the 21a or 21c type). This feature is observed in I_2 gas [20] when ω_1 is shifted slightly about ω_{na}. It facilitates the assignment of the lines. Further, the observation of double-electronic resonances is of considerable importance to the spectroscopist since one can easily interpret the corresponding $\omega_{n'a}$ absorption lines if the ω_{na} transition in resonance with ω_1 is known. This can facilitate the analysis of unknown portions of absorption spectra. A set of handy notations has been introduced in order to label the double or triple resonances. The three molecular transitions appearing in the Raman, laser and anti-Stokes denominators of the $\rho_{aa}^{(0)}$ term of Eq. (15) are listed between brackets, in that order; the J quantum number of $|a\rangle$ is indicated to the right of the last bracket.

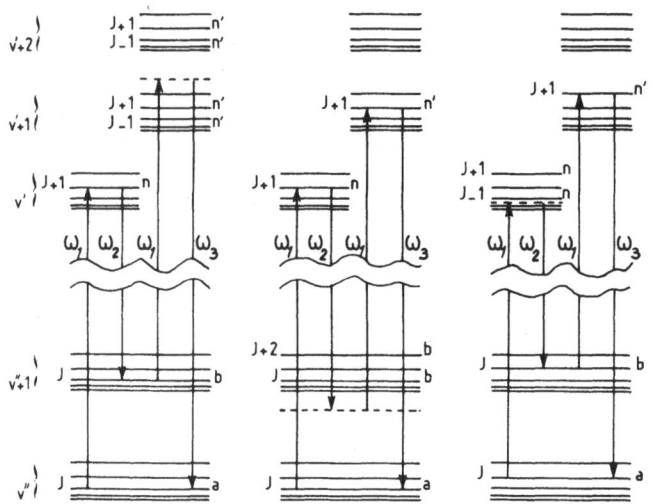

Fig. 21. Energy level diagrams representing the states contributing
          to resonance-enhanced CARS in a diatomic molecule:
          a) ω_{na} enhanced fundamental ($\Delta v = 1$) vibrational transition;
          b) double-electronic resonance with $\omega_1 = \omega_{na}$ and $\omega_3 = \omega_{n'a}$;
          c) $\omega_{n'a}$ enhanced vibrational transition.

Capital letters are used for the transitions in resonance with the
fields, lower case letters otherwise. Recently, resonance-enhanced
CARS spectra of C_2 in a discharge have been obtained under controlled
conditions by precisely tuning the "laser" and "Stokes" frequencies
into resonance with Swan band absorption lines and have been interpreted unambiguously [54].

   For that interpretation, we had to obtain the spectral positions
of the Swan emission lines to a precision of better than 0.01 cm^{-1}
using Fourier transform spectroscopy; this work was carried out in
collaboration with P. Luc and C. Amiot at CNRS [55].

   We first undertook the recording of the emission spectrum of
C_2. A stable microwave discharge source for the Swan band was
developed for that purpose. With a spectral resolution of 0.07 cm^{-1},
the characteristic triplet splitting and the staggering of the
triplet components caused by the Λ doubling are clearly visible in
the spectra. The latter study has provided a thorough and accurate
set of emission frequencies together with the appropriate set of
spectroscopic constants which were not available from earlier work
[56]. If v, J and Ω are respectively the vibrational, rotational
and spin quantum numbers, an a priori calculation shows that a
nearly triple resonance is achieved for one type of resonance CARS
line which is (in the notation or ref. [3]): $[Q_\Omega(1-0), R_\Omega(0-0),
P_\Omega(1-0)]J$ with J = 18, 19 and 20, for Ω = 3, 2 and 1 respectively.

These C_2 CARS lines have been searched for and observed by means of a microwave discharge in a flowing mixture of 3% acetylene in He at 40 mb total pressure. The "laser" and "Stokes" beams are focused into the discharge by a 30 cm focal length lens. The anti-Stokes signal is filtered by a monochromator and detected by a photomultiplier. Fluorescence in Iodine was used to calibrate the laser frequency ω_L; final accuracy is about 0.05 cm^{-1}. The spectrum of Figure 22 has been recorded with ω_L = 19493.85 ± 0.05 cm^{-1}.

Because of the high vibrational temperature, additional contributions are involved in the resonant CARS susceptibility. These are associated with the terms proportional to $\rho_{bb}^{(0)}$ of Eq. (15). Thus our CARS spectra present previously unobserved characteristics, particularly with regard to spectral content. Resonance CARS lines of the type $[(Q_1(1-0), R_1(0-1), P_1(1-0))]20$, $[(Q_2(1-0), R_2(0-0), P_2(1-0))]19$ and $[(Q_3(1-0), R_3(0-0), P_3(1-0))]18$ are seen. Different features of the type $[(Q_1(1-0), R_1(0-1), P_1(1-0))]20$, $[(Q_2(1-0), R_2(0-1), P_2(1-0))]19$ and $[(Q_3(1-0), R_3(0-1), P_3(1-0))]18$ are probably mixed to the above ones. All these lines are lumped together in the profile of Figure 22. Their widths are about 0.12 cm^{-1} and contain Doppler and collisional contributions.

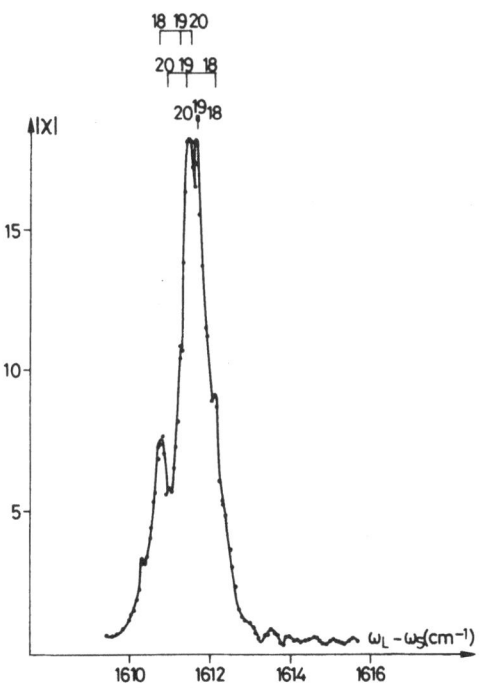

Fig. 22.   Resonance CARS spectrum of C_2 in a discharge.

2. Resonance CARS in the Doppler Regime

Resonance CARS in dilute samples, where the Doppler effect is the main cause of line broadening, is a problem of considerable interest. An intuitive, but fallacious argument, leads to the erroneous conclusion that resonance CARS lines should be Doppler-free. We take, for example, a laser-enhanced Raman resonance (Figure 21a). One particular velocity group in the Doppler-broadened ω_{na} absorption contour is precisely resonant; we call v its velocity component along the wave vector axis $v = c(\omega_{na}-\omega_1)/\omega_1$. One expects this group to give a dominant contribution in the CARS spectrum as $\omega_1 - \omega_2$ is being swept past ω_{ba}, thus giving a Doppler-free resonance line, which is shifted from the Raman resonance line center by the amount $\omega_{ba}v/c$. This simple-minded argument actually turns out to be wrong, because we have neglected the influence of adjacent velocity groups which, in this case, interfere destructively with the main group [23].

The correct approach is to recast Eq. (15) into its Doppler form. The molecules in the above-mentioned velocity group see the laser and anti-Stokes frequencies with a Doppler shift $-k_iv$, where k_i is the wave vector. Their elementary contribution to the susceptibility is shown to be

$$X_r(v) = \frac{\rho_{aa}^{(0)}}{h^3} \sum_b \frac{1}{\omega_{ba}-\omega_1+\omega_2+(k_1-k_2)v-i\Gamma_{ba}}$$

$$x \sum_{n'} \frac{\mu_{an'}\mu_{n'b}}{\omega_{n'a}-\omega_3+k_3v-i\Gamma_{n'a}} \qquad (27)$$

$$x \sum_n \frac{\mu_{bn}\mu_{na}}{\omega_{na}-\omega_1+k_1v-i\Gamma_{na}}$$

where we have neglected the $\rho_{bb}^{(0)}$ contribution. The net susceptibility is finally obtained by integrating over the velocity distribution function $F(v)$:

$$X_r = \int_\infty^\infty NF(v) \ X_r(v)dv \qquad (28)$$

Equation (27) can be integrated numerically. A closed form solution in terms of complex error functions is also easily derived if $F(v)$ is taken as a Gaussian. In this manner, it is possible to show that the $\rho_{aa}^{(0)}$ term in Eq. (15) indeed has the full Doppler width, whereas the $\rho_{bb}^{(0)}$ term is Doppler-free. Therefore, the latter may eventually give rise to the strongest lines in the spectrum under certain thermodynamic conditions, although $\rho_{bb}^{(0)} < \rho_{aa}^{(0)}$. A few of the other susceptibility terms in CARS are also Doppler-free; some contain at least one anti-resonance and should be small; others contain a resonance of the type $\omega_{n'n} - \omega_1 + \omega_2$, which under certain conditions on the collision rates can be quite large [17,24] (see discussion at the end of 2b).

In resonant CSRS, the Raman-resonant term which dominates in the collision regime is Doppler-free. We have also investigated the Doppler nature of the susceptibility terms in Stimulated Raman Gain Spectroscopy (SRGS) under a variety of experimental situations (co or counter-propagating laser waves, and Stokes or inverse Raman gain measurements). Depending on these conditions, a large number of terms can become Doppler-free (ref. 23, addendum). Note that a partial discussion of this same problem had already been given by Hansch and Toschek [57]. In short, we conclude from this analysis that CSRS and SRGS in principle are better suited than CARS for Doppler-free spectroscropy; however, we must bear in mind that they are prone to fluorescence interference.

5. CONCLUSION

We have examined in detail the theory of CARS, we have presented the key instrumental requirements and have reviewed some experimental results obtained to date. In brief, CARS:

1)   gives the same spectroscopic information as spontaneous Raman scattering;
2)   is insensitive to fluoresence interference in resonance Raman work;
3)   gives excellent spatial resolution (1 mm);
4)   can give good time resolution (1C nsec) by sacrificing some detectivity;
5)   is capable of excellent spectral resolution (0.03 cm^{-1} on ordinary set-ups, 10^{-3} cm^{-1} in special applications) using cw sources;
6)   is extremely luminous.

CARS has a few disadvantages:

1)   the major disadvantage is the presence of the non-resonant background, which limits the detectivity at a value of 10 ppm to 1% depending on thermodynamic conditions and species studied;
2)   it is sensitive to laser instabilities;
3)   it is subject to saturation at the higher power levels (1 MW or more).

For these reasons, CARS will often be preferred to normal Raman Scattering for the following measurements:

1)   study of reactive media, plasmas, gas laser amplifiers, etc...;

2)   analysis of media containing particulate matter (e.g., sooting flames) or investigation of flows near solid obstacles (as close as 50 µm to the surface);
3)   high resolution spectroscopy;
4)   resonance-enhanced Raman spectroscopy of liquids and gases.

Due to the cost of a state of the art CARS set-up (200,000 to 300,000$ for flame diagnostics), the decision to prefer CARS to normal Raman should be carefully weighed. However, fundamental research in all these areas will remain one of the most active branches of molecular spectroscopy.

REFERENCES

1.   G. F. Widhopf and S. Lederman, AIAA J., 9:309 (1971).
2.   M. Lapp, L. M. Goldman and C. M. Penney, Science, 175:1112 (1972).
3.   "Laser Raman Gas Diagnostics", Proceedings of the Project SQUID Laser Raman Workshop on the Measurement of Gas Properties, May 10-11, 1973, Schenectady, eds., M. Lapp and C. M. Penney, Plenum Press, New York, London (1974).
4.   Proceedings of Project SQUID Workshop on Combustion Measurements in Jet Propulsion Systems, ed., R. Goulard, Purdue University, Lafayette, Indiana (1975).
5.   "Experimental Diagnostics in Gas Phase Combustion Systems", Progress in Astronautics and Aeronautics, Vol. 53, ed., B. T. Zinn, Martin Summerfield Series Editor (1977).
6.   P. R. Régnier and J. P. E. Taran, Appl. Phys. Letters, 23:240 (1973).
7.   R. W. Terhune, Bull. Amer. Phys. Soc., 8:359 (1963).
8.   P. D. Maker and R. W. Terhune, Phys. Rev., 137:A801 (1965).
9.   E. Yablonovitvh, N. Bloembergen and J. J. Wynne, Phys. Rev., B3:2060 (1971).
10. S. A. Akhamanov, V. G. Dmitriev, A. I. Kovrigin, N. I. Koroteev, V. G. Tunkin and A. I. Kholodnykh, JETP Letters, 15:425 (1972).
11. M. D. Levenson, C. Flytzanis and N. Bloembergen, Phys. Rev., 6:B3962 (1972).
12. W. G. Rado, Appl. Phys. Letters, 11:123 (1967).
13. G. Hauchecorne, F. Kerhervé and G. Mayer, J. de Physique, 32:47 (1971).
14. F. De Martini, G. P. Giuliani and E. Santamato, Optics Comm., 5:126 (1972).
15. J. Fiutak and J. Van Kranendonk, Can. J. Phys., 40:1085 (1962).
16. A. Omont, E. W. Smith and J. Cooper, The Astroph J., 175:185 (1972).
17. S. Druet, B. Attal, T. K. Gustafson and J. P. E. Taran, Phys. Rev., A18:1529 (1978).
18. S. I. Yee, T. K. Gustafson, S. A. J. Druet and J. P. E. Taran, Opt. Commun., 23:1 (1977).

19. S. Y. Yee and T. K. Gustafson, Phys. Rev., A18:1597 (1978).
20. S. A. J. Druet and J. P. E. Taran, "Coherent anti-Stokes Raman
     Spectroscopy", in: Chemical and Biochemical Applications of
     Lasers, vol. 4, ed., C. B. Moore, Academic Press, New York
     (1979).
21. C. J. Bordé, J. L. Hall, C. V. Kunasz and D. G. Hummer, Phys.
     Rev., A14:236 (1976).
22. J. Bordé and C. J. Bordé, J. Mol. Spectrosc., 78:3530 (1979).
23. S. A. J. Druet, J. P. E. Taran and C. J. Bordé, J. de Physique,
     40:819 (1979), addendum, ibidem, 41:183 (1980).
24. S. A. J. Druet and J. P. E. Taran, Progress in Quant. Elec.,
     7:1 (1981).
25. R. P. Feynman, "Quantum Electrodynamics", Benjamin, New York
     (1962).
26. J. F. Ward, Rev. Mod. Phys., 37:1 (1965).
27. A. Yariv, IEEE J. Quant. Elect., QE13:943 (1977).
28. D. C. Hanna, D. Cotter and M. Yuratich, "Non-linear Optics of
     Free Atoms and Molecules", Springer Series in Optical
     Sciences, Vol. 17, ed., D. L. McAdam, Springer Verlag, Berlin,
     Heidelberg, New York (1979).
29. G. L. Eesley, J.Q.S.R.T., 22:507 (1979).
30. N. Bloembergen, H. Lotem and R. T. Lynch, Indian J. Pure Appl.
     Phys., 16:151 (1978).
31. Y. Prior, A. R. Bogdan, M. Dagenais and N. Bloembergen, Phys.
     Rev. Letters, 46:111 (1981).
32. A. R. Bogdan, Y. Prior and N. Bloembergen, Opt. Letters, 6:82
     (1981).
33. N. Bloembergen, A. R. Bogdan and M. W. Downer, in: "Laser
     Spectroscopy V", eds., McKellar, Oka and Stoicheff, Springer
     Verlag, Berlin, Heidelberg, New York (1981).
34. L. A. Carreira, L. P. Gross and T. B. Malloy, J. Chem. Phys.,
     69:855 (1978).
35. M. Grynberg, J. Phys. B. Atom Mol. Phys., 14:2089 (1981).
36. A. C. Eckbreth, Appl. Phys. Letters, 32:421 (1978).
37. L. A. Rahn, L. J. Zych and P. L. Mattern, Opt. Comm., 30:249
     (1979).
38. W. B. Roh, P. W. Schreiber and J. P. E. Taran, Appl. Phys.
     Letters, 29:174 (1976).
39. B. Attal, M. Péalat and J. P. E. Taran, J. Energy, 4:135 (1980).
40. J. J. Song, G. L. Eesley and M. D. Levenson, Appl. Phys.
     Letters, 29:567 (1976).
41. P. R. Régnier and J. P. E. Taran, Appl. Phys. Letters, 23:240
     (1973).
42. M. Péalat, J. P. E. Taran, J. Taillet, M. Bacal and A. M.
     Bruneteau, J. Appl. Phys., 52:2687 (1981).
43. M. Bacal and G. W. Hamilton, Phys. Rev. Letters, 42:1538 (1979).
44. J. W. Nibler, W. M. Shaub, J. R. McDonald and A. B. Harvey,
     "Coherent Anti-Stokes Raman Spectroscopy", in: Vibrational
     Spectra and Structure, Vol. 6, ed., J. R. Durig, Elsevier,
     Amsterdam, New York (1977).

45. K. Müller-Dethlefs, M. Péalat and J. P. E. Taran, <u>Ber.</u>
     <u>Bunsenges, Phys. Chem.</u>, 85:803 (1981).
46. J. Flossodorf, W. Jost and H. Cg. Wagner, <u>Ber. Bunsenges, Phys.</u>
     <u>Chem.</u>, 78:378 (1974).
47. P. Bouchardy, P. Gicquel, M. Péalat and J. P. E. Taran, to be
     published.
48. J. Bédué, P. Gastebois, R. Bailly, M. Péalat and J. P. E. Taran,
     to be published.
49. K. P. Gross, D. M. Guthals and J. W. Nibler, <u>J. Chem. Phys.</u>,
     70:4673 (1979).
50. W. M. Hetherington III, G. M. Korenowski and K. B. Eisenthal,
     <u>Chem. Phys. Letters</u>, 77:275 (1981).
51. D. M. Guthals, K. P. Gross and J. W. Nibler, <u>J. Chem. Phys.</u>,
     70:2393 (1979).
52. M. E. McIlwain and J. C. Hindman, <u>J. Chem. Phys.</u>, 73:68 (1980).
53. A. Beckman, H. Fietz, P. Baierl and W. Kiefer, <u>Chem. Phys.</u>
     <u>Letters</u>, 86:140 (1982).
54. B. Attal, K. Müller-Dethlefs, D. Débarre and J. P. E. Taran,
     to be published.
55. P. Luc, C. Amiot, D. Débarre, B. Attal and K. Müller-Dethlefs,
     "New Analysis of the C_2 Swan System", to be published.
56. J. G. Phillips and S. P. Davis, "The Berkeley Analysis of
     Molecular Spectra", Vol. 2(a), University of California
     Press (1968).
57. Th. Hansch and P. Toschek, <u>Z. Phys.</u>, 236:213 (1970).

REAL TIME TEMPERATURE AND CONCENTRATION MEASUREMENTS IN A

SEMI-INDUSTRIAL FURNACE WITH CARS SPECTROSCOPY

A. Ferrario and C. Malvicini

CISE SpA
Milano
Italy

INTRODUCTION

The use of Coherent Anti-Stokes Raman Spectroscopy (CARS) is undergoing rapid growth for non invasive measurements of gas temperature and composition in a broad range of practical combustion systems, such as internal combustion engines[1], bluff-body combustors[2] and turbine combustors.[3,4].

We report here the application of CARS techniques to a large scale ($3x3x6$ m^3) experimental section of a steel slab reheating furnace in order to get information about CARS spectrometer reliability and measurement accuracy in different operating conditions of the burners. We also report the preliminary laboratory measurements in a cell with gases at flame temperatures and the measurements on a smaller instrumented furnace carried out at the same time with conventional techniques.

In CARS two laser frequencies, ω_1 ("pump") and ω_2 ("Strokes"), with ($\omega_1-\omega_2$) tuned near a Raman active rotational or vibrational transition, interact through the third order non linear electrical susceptibility X^3 to generate an anti-Stokes shifted coherent wave at frequency $\omega_3=2\omega_1-\omega_2$. To maximize the conversion efficiency the wave vectors must comply with the phase-matching condition $\underline{k}_3=\underline{k}_1'+\underline{k}_1''-\underline{k}_2$ (Figure 1). Since the anti-Stokes signal generated from major species is collimated and often strong, it can be easily detected and discriminated against fluorescence and background flame radiation, even in many turbulent and sooting flames[5].

In virtually non dispersive gases collinear and focused laser beams give a phase-matched interaction with a spatial resolution of

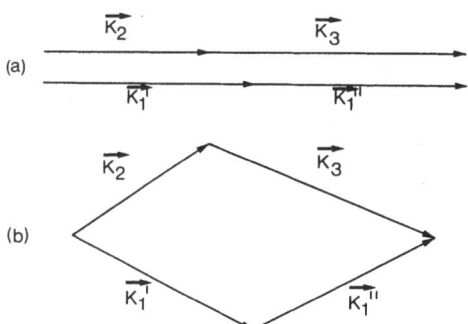

Fig. 1. Collinear (a) and BOXCARS (b) phase-matching diagrams.

about six times the confocal parameter[6]. To obtain a higher resolution the beams may be crossed in BOXCARS[7] phase-matching arrangement, which nevertheless gives a signal intensity 20-50 times lower.

The CARS spectra may be obtained by scanning a narrow band
Stokes laser and recording the anti-Stokes intensity versus $(\omega_1-\omega_2)$.
Another approach, termed "Multiplex CARS", may be utilized to obtain
CARS spectra of one or more molecules in a single laser shot, thus
overcoming the problems due to laser beam fluctuations and flame
turbulence. In multiplex CARS the anti-Stokes frequencies, generated
with a broad band Stokes laser, are dispersed on an optical multichannel analyzer.

Main drawbacks of CARS are the system complexity and a non
resonant signal presence due to electronic and off resonance Raman
transitions of all species that limits the detectivity at the 0.1-1%
level. This background signal may be strongly reduced by polarization techniques[4,6], with, however, increased complexity and
reduced useful signal. Nevertheless, for the direct measurement of
major species concentrations, it is possible to take advantage from
the presence of both resonant and non resonant signals in the same
spectrum.

THEORY

Since CARS theory is widely treated in literature[4,6,8], only a
simplified presentation is reported. The anti-Stokes intensity I_3
generated by phase-matched monochromatic beams is proportional to

$$|\chi^3|^2 \; I_1{}^2 I_2{}^2 \; L^2$$

where I_1 and I_2 are the laser intensities and L is the interaction
length. It can be shown that by focusing collinear gaussian laser
beams, the anti-Stokes generated power is nearly independent of the

focal length. The susceptibility can be expressed as

$$\chi^{(3)} = \chi^{NR} + \chi^{R} = \chi^{NR} + \Sigma_j \; \chi^{R}_j$$

where the real term χ^{R} is the non resonant contribution and the sum is made over the complex resonant susceptibilities χ^{R}_j due to near near resonance Raman transitions. The χ^{R}_j associated with a pressure broadened transition at frequency ω_j is proportional to

$$N \; \Delta_j \; (\frac{d\sigma}{d\Omega})_j \; \frac{1}{(\omega_j - \omega_1 + \omega_2) - i \; \Gamma_j/2}$$

where N is the number density of the probed molecules, Δ_j is the fractional population difference between the upper and lower vibration rotation level, $(\frac{d\sigma}{d\Omega})_j$ is the spontaneous Raman cross section and Γ_j is the Raman linewidth.

Actual spectra can be computed[9] as a proper convolution of the $|\chi^{(3)}|^2$ to-account for the finite pump laser linewidth and the optical multichannel analyzer resolution (or Stokes laser linewidth in scanning CARS experiments). If $G(\omega)$ is the overall spectrometer resolution function, the CARS spectrum may be written as

$$|\chi^{(3)}(\omega)|^2 \star G(\omega) = (\chi^{NR})^2 + 2 \; \chi^{NR} \; [\chi^{R}(\omega)' \star G(\omega)] +$$
$$+ [\; | \; \chi^{R}(\omega) \; |^2 \star G(\omega)]$$

where $\chi^{R}(\omega)'$ is the real part of resonant susceptibility and ω stands for the Raman shift $(\omega_1 - \omega_2)$.

Assuming that, at a constant temperature and total pressure, the Raman linewidths are independent of the relative gas concentrations, χ^{R} is proportional to the number density of resonant molecules. Hence, the ω-dependent functions between square brackets may be calculated for one relative concentration only (i.e. for pure gas).

Calculated spectra of N_2, O_2 and CO, at different concentrations and temperatures are shown in Figures 2-4, with 1.5 cm^{-1} spectral resolution.

CARS SPECTROMETER

The spectrometer we have realized for CARS spectroscopy essentially comprises a laser source, a polychromator, an optical multichannel analyzer and a minicomputer for data reduction (Figures 5 and 6).

The laser source has been mounted on a forecast aluminium table (Figure 7). At frequency ω_1 (5320 Å a frequency doubled Q-switched

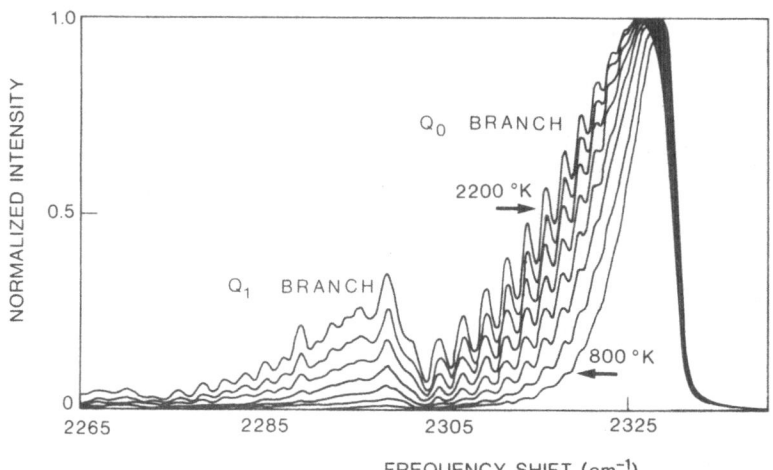

Fig. 2.   Computed N_2 spectra at a 78% concentration and different
          temperatures.

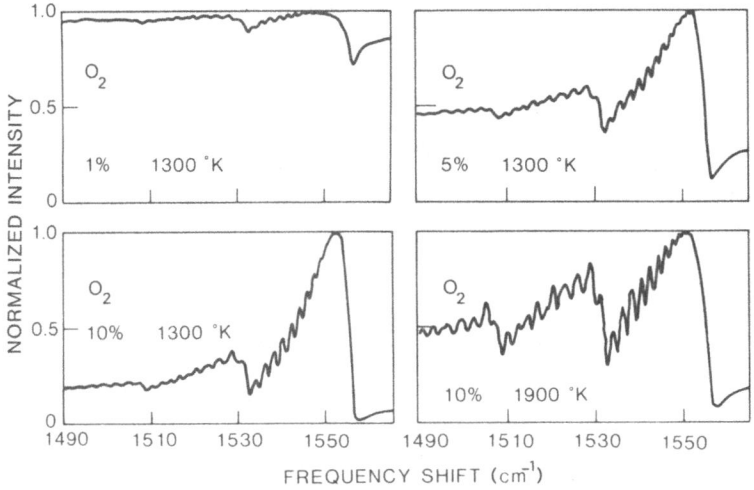

Fig. 3.   Computed O_2 spectra.

Nd-YAG laser, comprising a TEM_{OO} oscillator and two amplifiers,
delivers 150-200 mJ in 20 ns pulses up to 10 pps. The Stokes frequencies are obtained with a dye laser which is pumped quasi collinearly by one third of 5320 Å energy. The tuning range, 5600 - 6700 Å
with 1-10 mJ per pulse, is covered with different dye solutions,
which can be selected with solenoid valves.

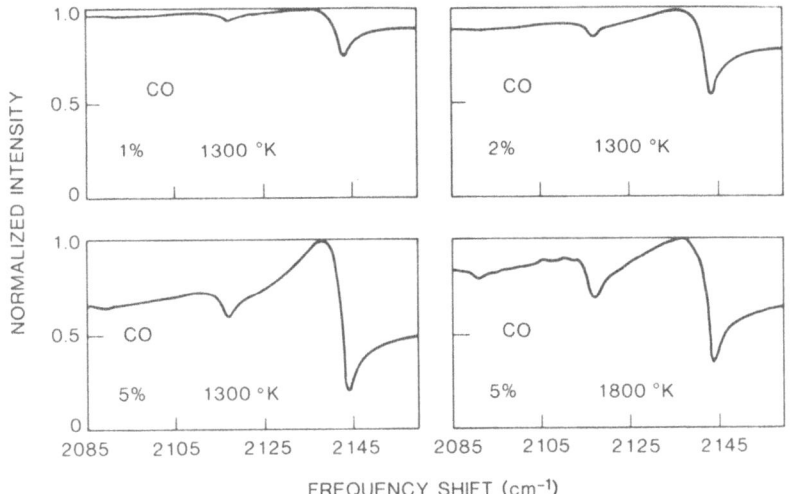

Fig. 4. Computed CO spectra.

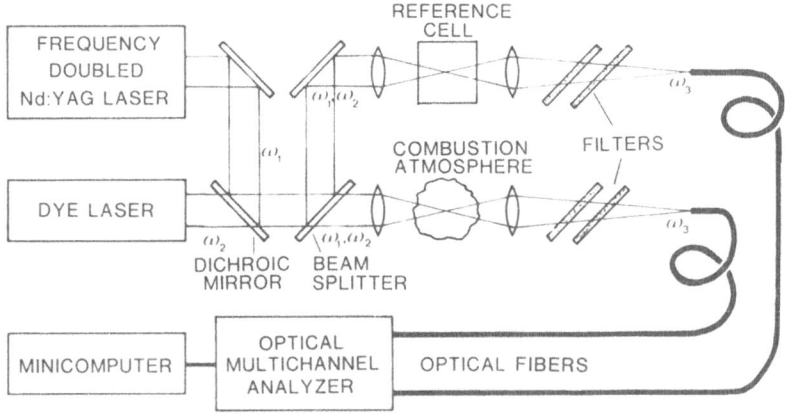

Fig. 5. Cars experimental setup.

The dye laser is usually operated in a broad band mode (100-150 cm^{-1} FWHM) for multiplex CARS, but it can also be operated with an 0,5 cm^{-1} band for scanning CARS experiments. In this case the cavity includes a grazing incidence fixed grating (2400 grooves/mm) and a tuning rotating mirror, driven by a stepping-motor. A thick plate can be inserted into the cavity for the laser switching between these two operating modes.

The two laser beams, with parallel polarizations, are super-imposed and aligned with a dichroic mirror, and then focused into the interaction medium. Optical components are provided to perform both

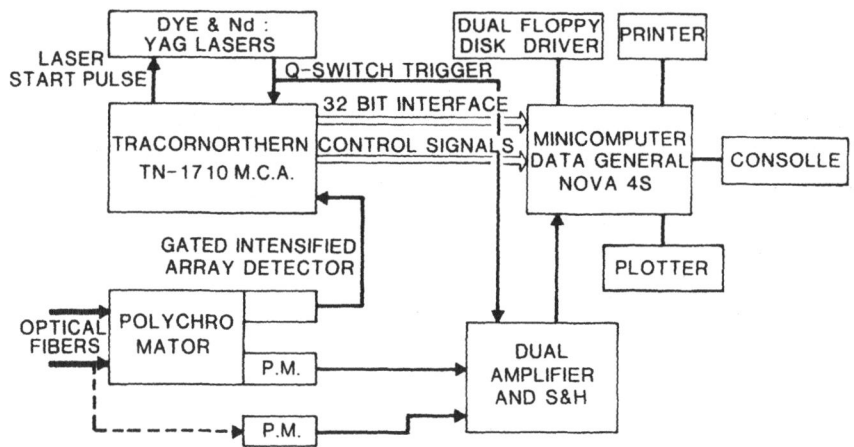

Fig. 6.   Block diagram of the CARS spectrometer.

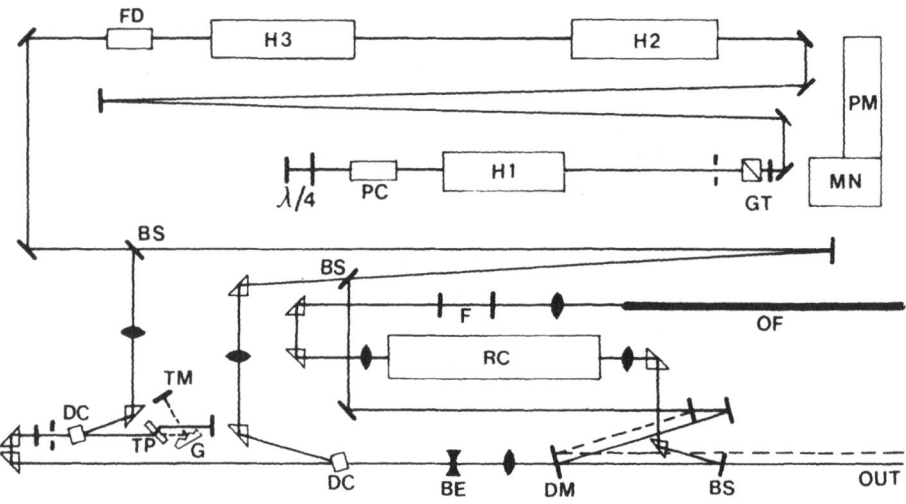

Fig. 7.   Laser source layout. PC:Pockels cell; H_1, H_2, H_3; Nd:YAG
          heads; GT: Glan Thompson polarizer; FD: KD*P frequency
          doubler; BS:beam splitter: DC:dye cells; G: grating; TM:
          tuning mirror; TP:thick plate; BE: beam-expander; DM:
          dichroic mirror. RC: reference cell; F: filters; OF:
          optical fiber; MN: tunable filter; PM: photomultiplier.
          The last two devices are utilized only for scanning CARS.

collinear and BOXCARS phase-matching. A small fraction of the laser
energy is split out and focused into a cell filled with argon gas at
30-40 atm to provide a non resonant signal for normalization
purposes.

Anti-Stokes signals generated in both legs are prefiltered and piped through two silica-silica optical fibers connected to two different entrance slits of an f/10.5 polychromator with one meter focal length, equipped with a 2400 grooves/mm holographic grating. A tiltable mirror is provided to select the desired slit. The dispersed anti-Stokes signals are detected by a gated intensified 1024 photodiode array, connected to a TRACTOR NORTHERN TN-1710 multichannel analyzer. The 25 μm spacing of the photodiodes corresponds to 0.09 Å and the FWHM resolution is about 0.3 Å.

The acquired spectra can be transferred to a DATA GENERAL NOVA 4S minicomputer through a 32 bit parallel interface. The time necessary for transferring one spectrum is about 140 ms. The spectra are then normalized ratioing them channel by channel with non resonant spectra obtained from the reference leg, in order to account for the shape of the dye laser spectral emission. The minicomputer has been programmed in Fortran language for real time (about 1 second) calculation of temperature and concentrations from N_2, CO_2, O_2, CO and H_2O spectra. Floppy disks are employed to record programs, data and spectra.

For scanning CARS experiments the detection apparatus consists of two photomultipliers and two gating integrators connected to the mini-computer.

SPECTRA ANALYSIS

The most accurate way to derive temperature and concentrations from the observed spectra is perhaps to perform a least-squares fit (best fit) with computed spectra. For a precise computation of spectra it is, however, important to have a good knowledge of Raman linewidths, transition frequencies, cross sections and non resonant susceptibilities.

Since these data are not completely available for all the molecules of interest, we have acquired BOXCARS experimental spectra in various mixtures of gases slowly flowing through a cell heated up to 1800°K by an oven. The temperature has been measured with a Pt-Pt/Rh 10% thermocouple. For N_2 spectra, frequently utilized for thermometry, we have found a satisfactory agreement (Figure 8) with spectra calculated utilizing the Raman linewidths given by Eckbreth[10]. Experimental CO and O_2 spectra in mixtures with N_2 at 1300°K allowed us to obtain some parameters, such as the ratio between Raman cross section and non resonant susceptibility so as to calculate the spectra with better accuracy.

The developed software for field measurements, instead of best fit routines, which are time and memory consuming, includes less accurate but faster and simpler spectra analysis routines. Experimental results are so displayed in an almost continuous mode on the system console.

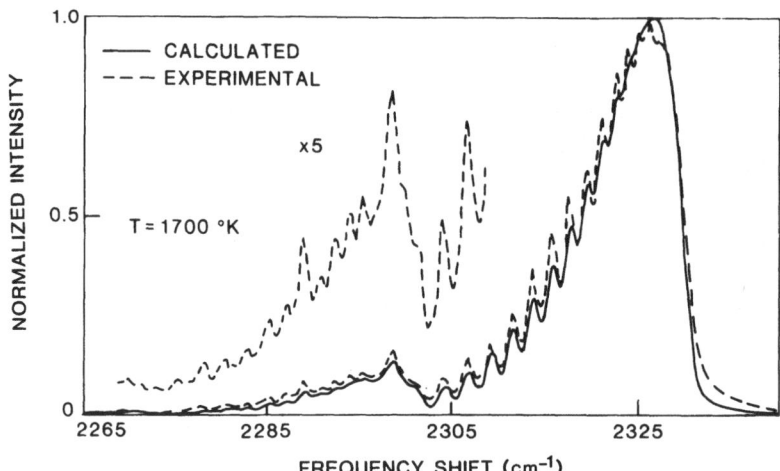

Fig. 8.   Comparison between an experimental and a computed N_2
         spectrum at 1700°K. In the experiment the temperature was
         measured with a thermocouple.

Thermometry is only performed on nitrogen, calculating the Q_1/Q_o
branch height ratio (see Figure 2). Its relation with the vibrational temperature (Figure 9) has been derived from computed spectra
and approximated with a polynomial expansion. Since for temperatures
lower than 1200°K the Q_1 band is barely resolved, the program also
calculates the area/height ratio of Q_o branch, which is related to
the rotational temperature.

For concentration measurements, two parameters are calculated
from experimental spectra:

$$m = \frac{(\text{maximum signal})^{\frac{1}{2}} - (\text{minimum signal})^{\frac{1}{2}}}{(\text{off resonance signal})^{\frac{1}{2}}}$$

$$s = \left(\frac{\text{resonant signal}}{\text{off resonance signal}} - 1\right)^{\frac{1}{2}}$$

It can be shown that for a Lorentzian line the parameter m is
proportional to the relative concentration[8]. In our experimental
CO, O_2 and H_2O spectra we calculate the parameter m on band heads
(see Figures 3,4). Since these spectra are not Lorentzian, the
linear relationship is valid only on a limited concentration range.
In fact it turns out either from laboratory measurements and from Co
and O_2 computed spectra at a 1300°K temperature that the linear
relationship is valid only for low concentrations, up to 2-3%. For

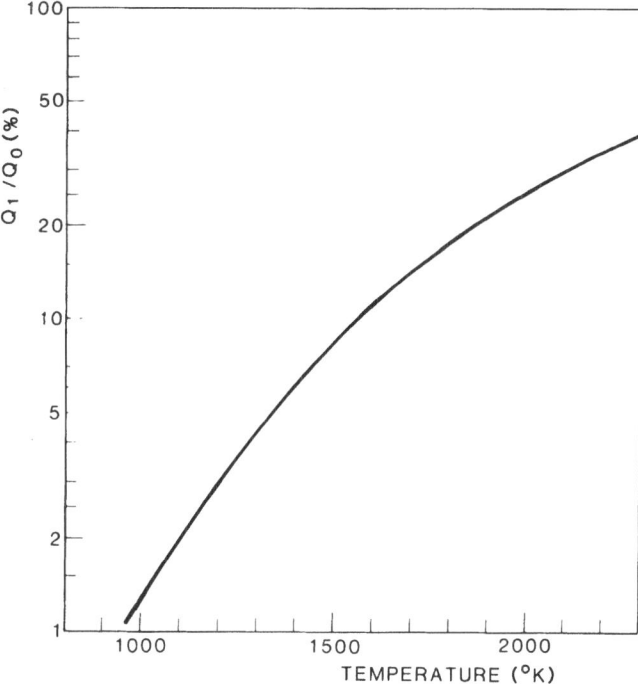

Fig. 9. Computed nitrogen Q_1/Q_0 branch height ratio versus
          temperature.

higher concentrations and for CO_2 spectra (Figure 10) the concentration is proportional to parameters. To calculate this parameter
an appropriate averaging around the Q_0 branch maximum value is made.

Either parameter m or s is chosen depending on molecules and
concentration ranges. Laboratory 1300°K experimental spectra have
been utilized for calibration. For concentration measurements at
different temperatures, the dependence of the two parameters on the
temperature has been calculated taking into account the population
variations of the vibration-rotation levels and assuming a $T^{-1/2}$
dependence of Raman linewidths. The χ^{NR} has been taken proportional
to the total number density, neglecting the dependence on the relative concentration of gases.

SMALL INSTRUMENTED FURNACE MEASUREMENTS

First field measurements were performed on a small instrumented
furnace, with an internal diameter of 40 cm and 90 cm length (Figure
11). The burner was fuelled with methane (up to 8 Nm^3/h) and air,
producing a 10 cm diameter gas flow. The CARS measurements were

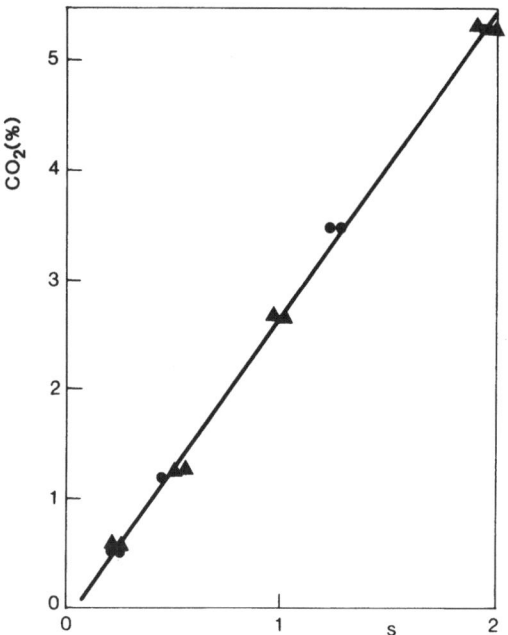

Fig. 10.   Experimental relationship between the CO_2 concentration
and the parameters s at 1300°K.   A 31 channels averaging
around the 1388 cm^{-1} band was made.

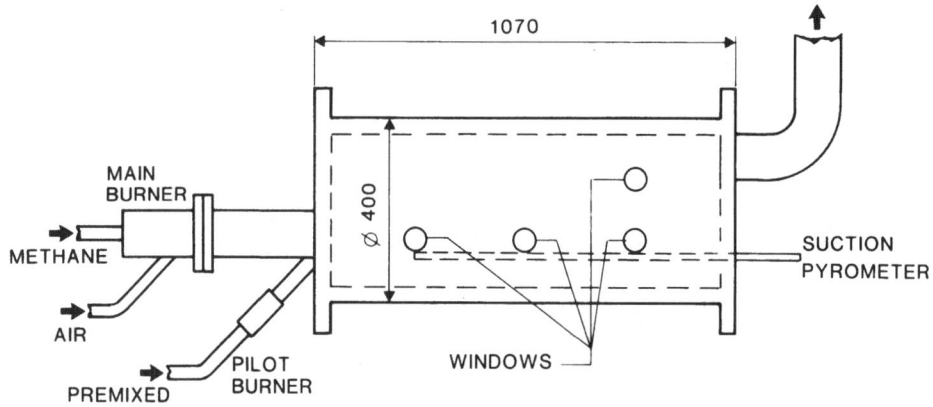

Fig. 11.   Small instrumented furnace.

usually performed in collinear phase-matching with a 70 cm focal
length lens, giving a $\sim$ 2 cm confocal parameter.   An appreciable CARS
signal was generated from cold N_2 and O_2 outside the furnace.   To
avoid this problem an OG515 filter was placed on the input window and
a dichroic mirror, reflecting the laser beams, was placed on the
output window.

Figure 12 shows a CARS temperature mapping along the laser beams, at the flame end. BOXCARS and suction thermocouple measurements performed in the hottest location are also shown. The discrepancy is mainly attributed to the limited spatial resolution of collinear phase-matching. CARS thermometry in the exhaust region, at 1300-1400°K, gave a satisfactory agreement with suction thermocouple measurements, within 20-30°K. The standard deviations were typically 20-30°K for 10 shots averaged and 70-80°K for single shot spectra. We think that they are almost only due to the dye laser mode competition and detection noise. CARS concentration measurements of CO_2, O_2 and H_2O were performed in the exhaust gas region. Typically the off resonance signal gave 10 to 100 photocounts per channel. Gas samples were also collected and analyzed with a gas chromatograph and with an ORSAT apparatus for CO, O_2 and CO_2 determination; water vapor was condensed and weighted. Figure 13 shows three spectra of CO_2 and O_2 obtained at different air/fuel ratios. When the samples were collected with proper care, the discrepancy between CARS and standard techniques was lower than 20%. Moreover, we were able to detect

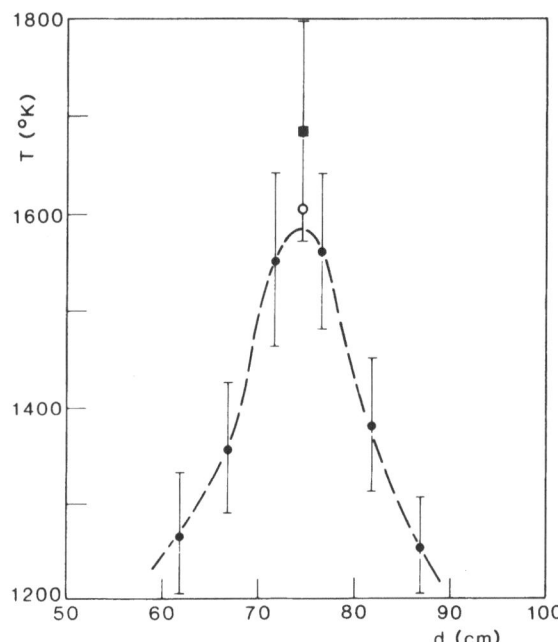

• SINGLE SHOT COLLINEAR CARS
■ SINGLE SHOT BOXCARS
o SUCTION PYROMETER

Fig. 12. Temperature mapping performed through the first window on the left side of Fig. 11. In each point 30 single shot measurements were averaged. The error bars represent ± one standard deviation.

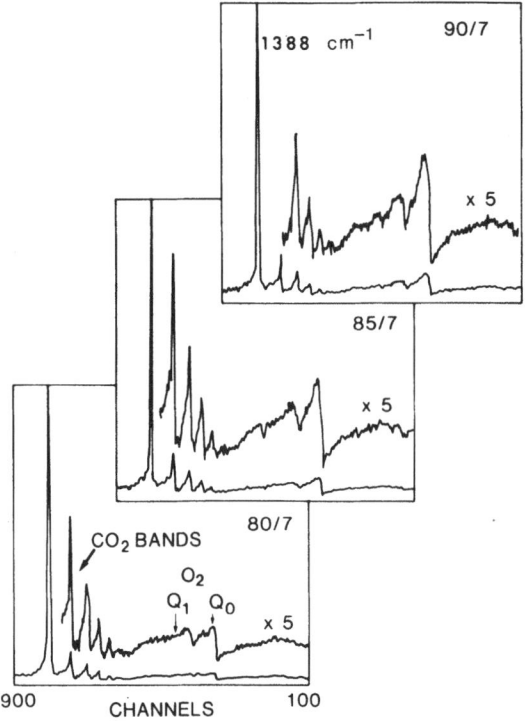

Fig. 13. Three CO_2 and O_2 spectra obtained in the small furnace (7
N m^3/h of methane and respectively 80,85 and 90 N m^3/h of
air).

spurious air input in the furnace from wide occasional fluctuations
of oxygen concentration measured with single shot spectra.

LARGE SCALE EXPERIMENTAL FURNACE MEASUREMENTS

     CARS measurements were performed on a large scale experimental
section of a steel slab reheating furnace (Figures 14 and 15), with
internal dimensions of 3x3x6 m^3. Two roof radiant burners fuelled
with methane (up to 100 Nm^3/h) and a lateral long-flame burner were
installed. The lateral burner had a 40 cm refractory air nozzle and
could be fuelled with methane (up to 150 Nm^3/h), gasoil or heavy fuel
oil (up to 80 1/h).

     Laser source, polychromator and electronics were housed in a
cabin. The laser beams crossed the lateral burner flame with more
than four meters path, including the refractory wall thickness,
inside the furnace. Two 50 mm diameter windows were installed, with
nitrogen flow to keep them clear. Optical filters were provided on

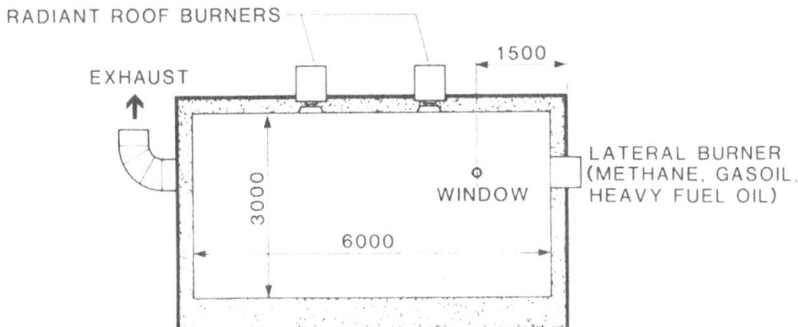

Fig. 14. Simplified section of the large scale furnace.

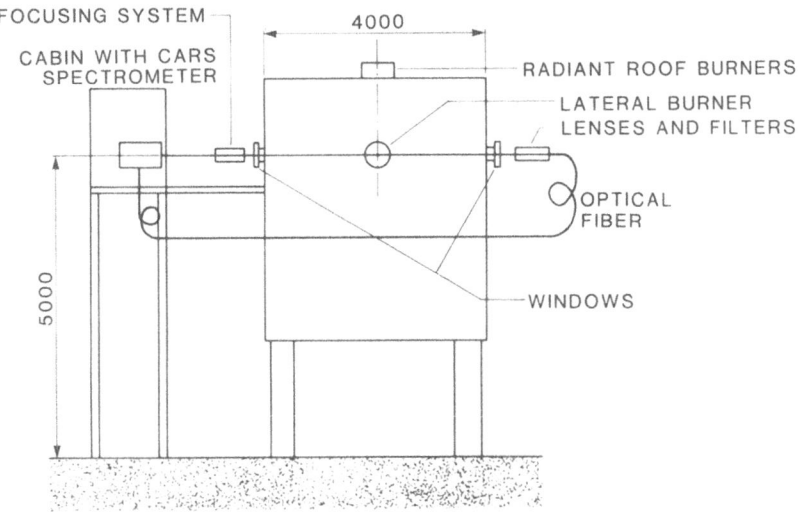

Fig. 15. Experimental setup for CARS measurements in the large
          scale furnace.

the windows to avoid anti-Stokes signal generation outside the
furnace.

A three lenses focusing system (Figure 16) was utilized in order
to easily move the beam-waist location along the laser beam axis,
while maintaining a constant 10 cm confocal parameter. It consisted
of a 50 cm focal length lens, followed by a fixed beam-expander with
magnification M = 3.3 A 25 meters long silica-silica optical fiber
(core $\emptyset$ = 200 μm, N.A. = 0.13) piped the anti-Stokes signal from the
output window to the cabin. This optical fiber was chosen after a
measurement of He-Ne laser beam wandering at the furnace outlet.

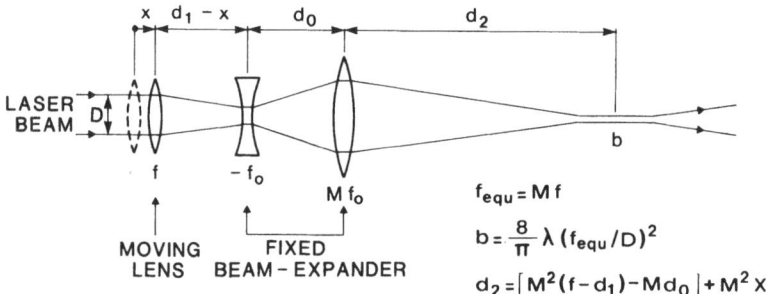

$$f_{equ} = M f$$

$$b = \frac{8}{\pi} \lambda (f_{equ}/D)^2$$

$$d_2 = \left[M^2(f-d_1) - M d_0 \right] + M^2 x$$

Fig. 16. Focusing system.

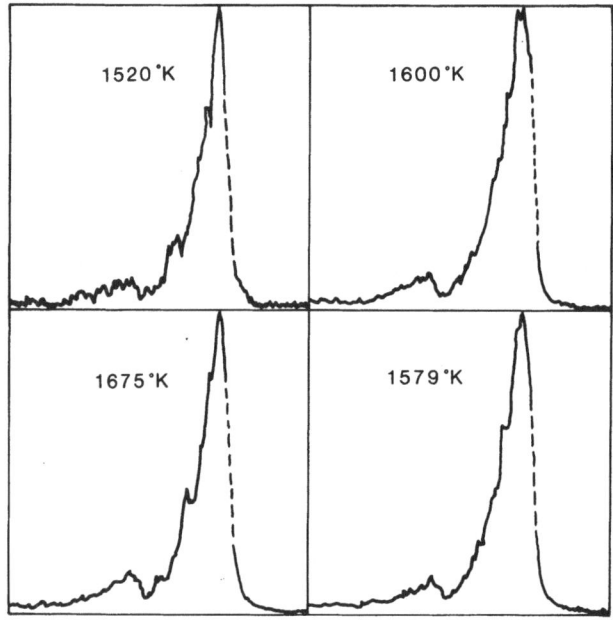

Fig. 17. Four successive single shot spectra of nitrogen in gasoil
          flame. The calculated temperatures are also shown.

Highly sooting and luminous gasoil flames were tested. The
comparison with the small furnace experiments showed that the signal
strengths and the standard deviation of the measurements were practically the same. Figure 17 shows four successive single shot nitrogen
spectra obtained with gasoil. The calculated temperatures are also
shown. Occasionally, the C_2 Swan band was noticed in the tail of
nitrogen Q_0 branch. The temperature value, calculated from Q_1/Q_0

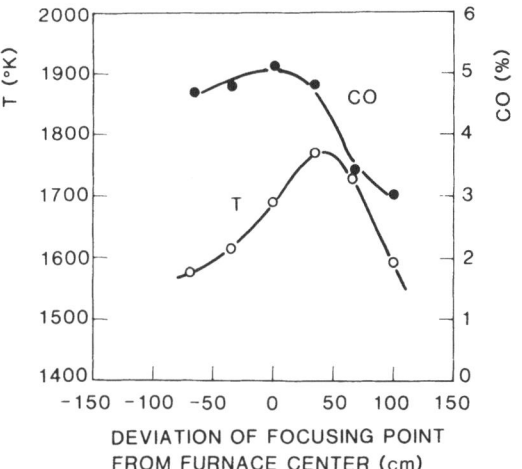

100 LASER SHOTS PER POINT
LATERAL BURNER: 780 Nm3/h AIR
                +120 Nm3/h METHANE
ROOF BURNERS : 400 Nm3/h AIR
                +66 Nm3/h METHANE
WALLS TEMPERATURE: ABOUT 1400 °K

Fig. 18. Simultaneous temperature and CO concentration mapping in
          the large scale experimental furnace.

branch height ratio was, however, not affected. A mapping of temperature and CO concentration, with methane fuel, is reported in
Figure 18. Typical H_2O and $CO-N_2$ spectra are shown in Figures 19

and 20. With gasoil fuel occasional (10-20% of shots) interferences,
due to hydrogen and hydrocarbons, were noticed in CO_2-O_2 spectra
(Figure 21).

CONCLUSIONS

     We have described CARS investigations in laboratory and real-
istic environments. A number of experimental results were accumu-
lated working also in a very large furnace equipped with burners fed
by various types of fuels. We have demonstrated that reliable operation of the CARS spectrometer is possible together with extraction
of real time, useful and precise quantitative temperature and concentration measurements also from large and highly sooting flames. Some
other investigations in other real environments are hopefully planned
in the future.

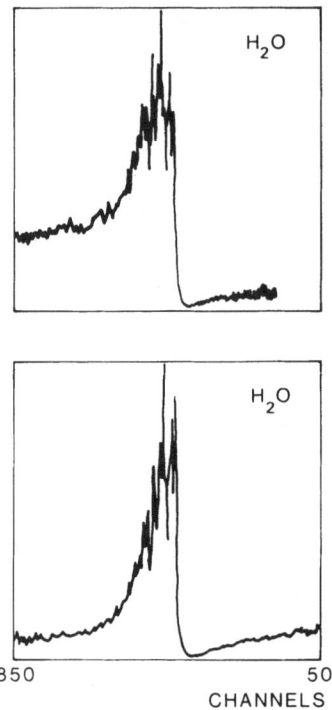

Fig. 19. Two typical water vapor spectra at different temperatures.

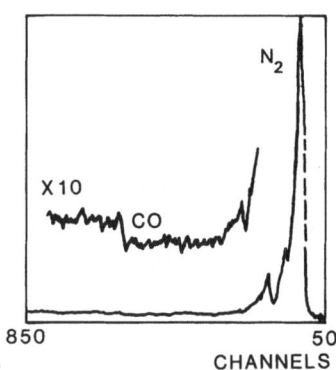

Fig. 20. Typical spectrum of CO and N_2.

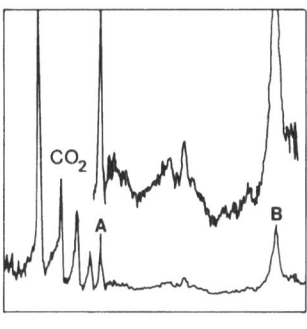

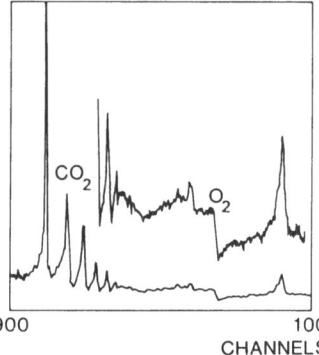

Fig. 21.   Spectra of CO_2 and O_2 in gasoil flame (respectively with
           fuel and with air excess).
           In the first one an hydrogen rotational line (A) adds to a
           CO_2 hot band. The (B) band is due to an hydrocarbon,
           probably ethylene (1623 cm^{-1}) or propylene (1620 cm^{-1}).
           Other interferences appear in the O_2 region.

REFERENCES

1.   D. Klick, K. A. Marko, and L. Rimai, Appl.Opt., 20:1178 (1981).
2.   G. L. Switzer, W. M. Roquemore, R. B. Bradley, P. W. Schreiber,
     and W. B. Roh, Appl.Opt., 18:2343 (1979).
3.   M. Péalat, J. P. Taran, and F. Moya, Optics and Laser Technology
     12:21 (1980).
4.   A. C. Eckbreth and P. W. Schreiber, in: "Chemical Applications
     of Nonlinear Raman Spectroscopy," A. B. Harvey, ed., Aca-
     demic Press, New York (1981).
5.   A. C. Eckbreth and R. J. Hall, Combust.Flame, 36:87 (1979).
6.   S. Druet and J. P. Taran, Progress in Quantum Electronics, 7:1
     (1981).
7.   J. A. Shirley, R. J. Hall, and A. C. Eckbreth, Opt.Lett., 5:380
     (1980).
8.   W. M. Tolles, J. W. Nibler, J. R. McDonald, and A. B. Harvey,
     Appl.Spectr., 31:253 (1977).
9.   A. J. Kotlar and J. A. Vanderhoff, Appl.Spectr., 36:421 (1982).
10. R. J. Hall, Appl.Spectr., 34:700 (1980).

LASER DETECTION OF COMBUSTION SPECIES

P. A. Bonczyk

United Technologies Research Center
East Hartford
Connecticut, USA

INTRODUCTION

Separate discussions are given of two laser-related in-situ diagnostic techniques pertinent to combustion media species characterization. The first is laser-induced fluorescence for determining the concentrations of selected molecules and/or radicals. The second is Mie scattering of laser radiation for determining the size and concentration of carbonaceous particulates. In discussing laser-induced fluorescence, attention will be given to: applicability criteria; simple two- and multi-level models of radiation interaction; the distinction between linear and saturated fluorescence; difficulties peculiar to combustion media; successful applications of the technique and its limitations. The discussion of Mie scattering will include: distinctions regarding size range, shape and composition of particulates; descriptions of scattering and extinction measurement schemes; the nature and relative importance of global and single particle scattering; problems particular to combustion media; measurement examples and technique limitations.

A. LASER-INDUCED FLUORESCENCE

There are a number of different atoms and molecules of interest in combustion processes. A detailed listing of them has been given by Eckbreth, Bonczyk and Verdieck[1]. The more important species include C, H, CH, NH, OH and even polyatomic radicals such as CH_2 and CH_3. For these and other species, there is a need for a non-perturbing, spatially and temporally precise method for determining their respective concentrations. In this regard laser-induced fluorescence is highly useful in that it may be made specific to the species of

interest and satisfies, as well, other requirements mentioned above.
The measurement procedure, in its simplest form, consists of popu-
lating the excited state of a molecule via the absorption of resonant
laser radiation, observing subsequent fluorescence emission from the
excited state, and inferring the concentration of ground state molecules present from the fluorescence intensity. The laser-induced
fluorescence technique is not applicable to all combustion species.
To be applicable, specific criteria must be satisfied. First the
atom or molecule must have a known emission spectrum. For a molecule
this in fact may not be the case if it deexcites via dissociation
prior to fluorescence emission. Second, the excitation wavelength
must be accessible to currently available tunable laser sources. At
present the wavelength range covered by these sources is roughly the
interval from 2000 Å to 1.5 μm. Atomic nitrogen, e.g. falls outside
this range since its first allowed transitions are deep in the vacuum
u.v. near 1200 Å. The third criterion is that the rate of radiative
decay of the excited state is known since the fluorescence power is
proportional to it. Fourth, in the presence of collisions of excited
state molecules with foreign species, the excited state decay rate
may significantly exceed the radiative rate. This collisional rate,
referred to as quenching, must be known as well. There is, however,
an important exception to this last requirement; namely, in the case
that saturation of the fluorescence occurs, recourse to quenching
data is not required. This will be evident from the discussions
which follow. Saturation requires intense pumping radiation, and
sets in when the ground and excited state populations approach equal-
ization.

The laser-induced fluorescence process for a two-level system is
shown in Figure 1. This representation is directly applicable to
selected atomic cases'. The upper level is populated by stimulated
absorption of resonant laser radiation, and deexcitation occurs via
stimulated emission, radiative decay and quenching. In Figure 1:
B_{12} and B_{21} are the Einstein coefficients for stimulated absorption
and emission, respectively; $I_{L\nu}$, the laser spectral intensity in
Watts/cm^2/cm^{-1}; A_{21}, the radiative decay rate; Q_{21}, the quenching
rate. The rate equations which govern the processes shown in Figure
1 are given by

$$dN_1/dt = -N_1 B_{12} I_{L\nu} + N_2 (A_{21} + Q_{21} + B_{21} I_{L\nu}) \qquad (1a)$$

and

$$dN_2/dt = + N_1 B_{12} I_{L\nu} - N_2 (A_{21} + Q_{21} + B_{21} I_{L\nu}), \qquad (1b)$$

where N_1 and N_2 are the populations of levels 1 and 2, respectively.
In a steady-state approximation where $dN_i/dt = 0$ ($i = 1,2$), it may be

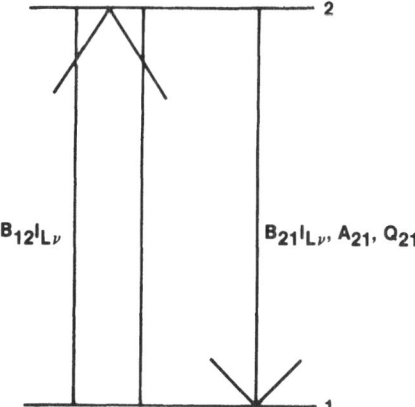

Fig. 1. Two-level laser induced fluorescence model.

shown that N_2 is related to the lower level population in the absence of the laser, $N_1{}^\circ$ $(= N_1 + N_2)$, by

$$N_2 = N_1{}^\circ B_{12} \left[\frac{Q_{21} + A_{21}}{I_{L_\nu}} + (B_{12} + B_{21}) \right]^{-1} \tag{2}$$

The fluorescence power S_F depends directly on N_2 and is given by

$$S_F = \frac{hc}{\lambda_F} \frac{A_{21}}{4\pi} \Omega V N_2, \tag{3}$$

where h is Planck's constant; c the speed of light; λ_F the fluorescence wavelength; Ω and V the light collection solid angle and sample volume size, respectively. Depending on the magnitude of I_{L_ν}, the fluorescence power may be linear or nonlinear (saturation) in its dependence on laser spectral intensity. In the limit of low laser intensity such that $I_{L_\nu} (B_{12} + B_{21})/(Q_{21} + A_{21}) \ll 1$, it follows from Equations (2) and (3) that

$$S_F{}^{(1)} = \frac{hc}{\lambda_F} \frac{B_{12}}{4\pi} \Omega V \left[\frac{A_{21}}{Q_{21} + A_{21}} \right] I_{L_\nu} N_1{}^\circ. \tag{4}$$

In this case $S_F{}^{(1)}$ depends linearly on I_{L_ν} and on the quenching rate Q_{21} through the factor $[A_{21}/(Q_{21} + A_{21})]$. In the other limit of high laser intensity such that $I_{L_\nu} (B_{12} + B_{21})/(Q_{21} + A_{21}) \gg 1$, it follows also from Equations (2) and (3) that

$$S_F^{(2)} = \frac{hc}{\lambda_F} \frac{A_{21}}{4\pi} \ \Omega V \left[1 + \frac{g_1}{g_2} \right]^{-1} N_1^0, \tag{5}$$

where the relation $(B_{21}/B_{12}) = g_1/g_2$ has been used in which g_1 and g_2 are level degeneracies. The power $S_F^{(2)}$ corresponds to full saturation of the emission, and is independent of both $I_{L\nu}$ and Q_{21}. The significance of the preceding inequality is that saturation occurs when the stimulated rates significantly exceed the combined quenching and radiative rates. Importantly full saturation obviates the need to know Q_{21} in order to evaluate N_1^0, as is evident from Equation (5) above. In combustion applications, Q_{21} is often very large and, hence, full saturation is difficult if not impossible to achieve. In the case of near saturation, which often is accessible, Baronavski and McDonald[2] have discussed an approach which circumvents the need for an a priori Q_{21} value. For near saturation, the fluorescence power is given approximately by

$$S_F^{(3)} \simeq \frac{hc}{\lambda_F} \frac{A_{21}}{8\pi} \ \Omega V N_1^0 \left[1 - \frac{(Q_{21} + A_{21})}{(B_{21} + B_{12}) I_{L\nu}} \right] . \tag{6}$$

In Equation (6), $S_F^{(3)}$ has a linear dependence on $(I_{L\nu})^{-1}$, if $S_F^{(3)}$ is plotted versus $(I_{L\nu})^{-1}$, the intercept on the $S_F^{(3)}$ axis is proportional to N_1^0, and the slope of this straight line is proportional to $(Q_{21}N_1^0)$, permitting separate determination of both N_1^0 and Q_{21}.

The preceding description of concentration measurement via laser-induced fluorescence is appropriate for atoms. A modified approach is required for molecules in view of their multi-level properties. An approximate treatment of the molecular case has been given by Verdieck and Bonczyk[3]. The interaction of laser radiation with a molecule having electronic, vibrational, and rotational degrees of freedom is shown schematically in Figure 2. The two larger rectangles represent unperturbed rotational levels of vibrational states having different electronic energies. A fraction f of the population N_1 of the lower group of levels is perturbed at a rate $B_{12}I_{L\nu}$, by a laser interacting with one rotational level. This occurs since rotational relaxation is fast, cross-relaxation occurs and other levels participate. The total number of participating levels is N_1f. In the limit of very slow relaxation, f is the rotational Boltzmann factor, whereas f approaches unity for very fast relaxation. Using a rate-equation approach similar to Equations (1a)

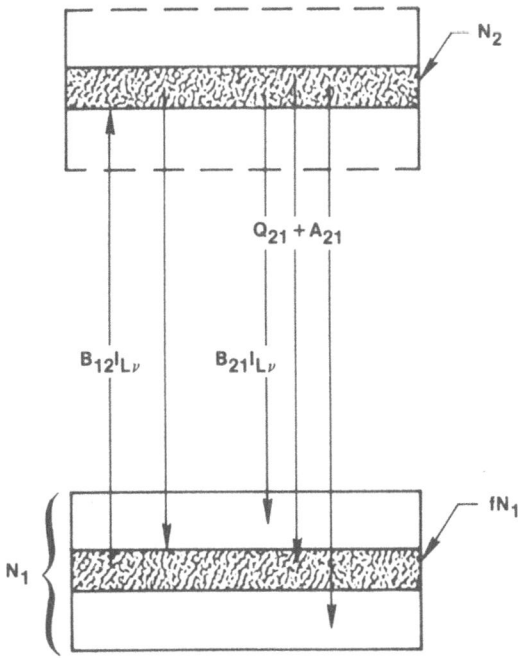

Fig. 2.   Modified two-level model for saturated fluorescence with
          rotational redistribution. [Reprinted from Reference 3 with
          the permission of the Combustion Institute].

and (1b) above, and making a steady-state approximation, it may be
shown that

$$N_2 = \frac{fN_1{}^o}{\left[f + \dfrac{g_1}{g_2} \right]} \left[1 + \frac{(Q_{21} + A_{21})}{\left[f + \dfrac{g_1}{g_2} \right] B_{12} I_{L_\nu}} \right]^{-1} \tag{7}$$

The fluorescence power for "near saturation" is given approximately
by

$$S_F \simeq \frac{hc}{\lambda_F} \frac{A_{21}}{4\pi} \frac{\Omega V N_1{}^o f}{\left[f + \dfrac{g_1}{g_2} \right]} \left[1 - \frac{(Q_{21} + A_{21})}{\left[f + \dfrac{g_1}{g_2} \right]^{B_{12}} I_{L_\nu}} \right] \tag{8}$$

Provided that f is known or may be approximated, $N_1{}^o$ and Q_{21} may be
determined by plotting S_F versus $I_{L_\nu}{}^{-1}$ similar to $S_F{}^{(3)}$ above. If f
= 1 and g_1/g_2 = 1, Equation (8) reduces to Equation (6) above. Other

models of laser-induced fluorescence incorporating rotational effects have been given by Berg and Shackleford[4] and Lucht et al.,[5].

Combustion Applications

There are several important factors particular to saturated laser-induced fluorescence measurements in combustion media. The first of these is the high rate of quenching Q_{21}, normally present at atmospheric pressure, which necessitates then a very high laser spectral intensity in order to achieve near saturation. For example, in the case of NO the quenching of the excited $A^2\Sigma$ state due to water vapor, which is a very efficient quenching agent, is 6.84×10^8 sec^{-1} Torr^{-1}. Typically, the partial pressure of water vapor in an atmospheric pressure flame is of order 120 Torr so that the quenching rate is, $Q_{21} = 8.2 \times 10^{10}$ sec^{-1}. For this rate, it may be shown that the calculated laser spectral intensity required to reach near saturation conditions is, $I_{L\nu} = 4 \times 10^8$ Watts/cm^2/cm^{-1}[1]. Saturated fluorescence has not been observed yet for NO. For CH in a premixed acetylene/oxygen flame[3], the measured quenching rate was, $Q_{21} = 8 \times 10^{11}$ sec^{-1}, for near saturation, i.e. $I_{L\nu} = 3 \times 10^8$ Watts/cm^2/cm^{-1}. Substituting appropriate numerical values for the focussed area and spectral width of the laser, 3.4×10^5 Watts of laser power were required for saturation. This level of power is not available from cw laser sources. Accordingly, a pulsed Nd:YAG pumped tunable dye laser was used for the work described in [Reference 3], and similar lasers are used in general for combustion measurements.

The principal sources of signal interference in laser-induced fluorescence combustion measurements are background luminosity and laser modulated particulate incandescence[1]. Background luminosity is comprised primarily of chemiluminescent emissions and the blackbody continuum from soot particulates. The former emissions are generally narrowband consisting of atomic line or molecular band structure specific to the emitting species and in distinct spectral regions. In some cases their interference can be avoided by appropriate laser wavelength selection. The blackbody continuum intensity usually exceeds chemiluminescence in a hydrocarbonfueled diffusion flame, whereas the reverse relationship is true for premixed flames. Laser modulated particulate incandescence is due to the absorption of laser light and the subsequent heating of already incandescent soot particulates to temperatures much higher than the ambient flame temperature with a resulting large increase in blackbody radiation. In general detailed signal/interference estimates are required in order to evaluate the relative importance of laser modulated and blackbody incandescence. In this regard it may be shown that laser modulated incandescence is the principal interference source in saturated fluorescence measurements utilizing high intensity laser sources. For saturation the fluorescence intensity is maximized since in steady-state the excited level of the species is heavily populated. The high laser intensities required, however, leads to increased particulate incandescence resulting in significant signal interference. It may be shown that this limit CH concentration measurements in flames to 80 ppm or higher[1].

In practical flames, the species concentrations may be both spatially and temporally inhomogeneous. There are methods of coping with the former, whereas temporal fluctuations occurring in turbulent flames are more difficult to handle. Spatial inhomogeneities require the fluorescence sample volume to be small enough so that the species concentration within it is constant. This, in fact, is possible in most cases by proper selection and arrangement of lenses, stops and apertures which constitute the light collection optics[6]. There is, however, a trade-off between spatial resolution and signal intensity in that the latter is proportional to the sample volume size. Successful CH saturated fluorescence measurements have been carried out[3] for a volume, $V = 5.7 \times 10^{-4}$ cm^3. In order to handle temporal fluctuations successfully, it is necessary to make single pulse saturated fluorescence measurements. This has not been achieved up to the present. Instead, time-averaged measurements have been made due to signal/interference limitations accompanying single pulse approaches[3]. There is one exception to this; namely, Stepowski and Cottereau[7] have made OH measurements from a single laser pulse. This measurement, however, has limited significance for practical flame applications in that it was done for a flat propane/oxygen flame operating at low, 20 Torr pressure.

Fluorescence Measurements

There have been several saturated laser-induced fluorescence measurements of species concentration in flames. Saturated molecular fluorescence has been observed for the species C_2,[8] CH,[3] CN,[3] MgO,[9] and OH[10]. In order to clarify the nature of such measurements, recent CH measurements are described here in some detail.

The apparatus used for CH measurements in Reference 3 is shown schematically in Figure 3. It consists of a laser-pumped tunable dye laser for producing radiation near 4300 Å, a burner for CH production, laser focussing and light collecting optics, and signal recording instrumentation. The pump laser in Figure 3 is a Nd:YAG laser having a 10 nsec pulse width and a 10 Hz pulse repetition frequency. The laser yields 200 mJ and 100 mJ of energy at 5320 and 3550 Å, respectively. The latter wavelengths are harmonics of the Nd:YAG 1.06-μ fundamental. The dye laser is comprised of oscillator, preamplifier, and final amplifier sections pumped by 3 x Nd:YAG radiation at 3550 Å. The oscillator dye cell is transversely pumped, and a prism/grating combination is used to tune and spectrally condense the oscillator output. A transversely pumped preamplifier and a longitudinally pumped final amplifier complete the dye laser. The

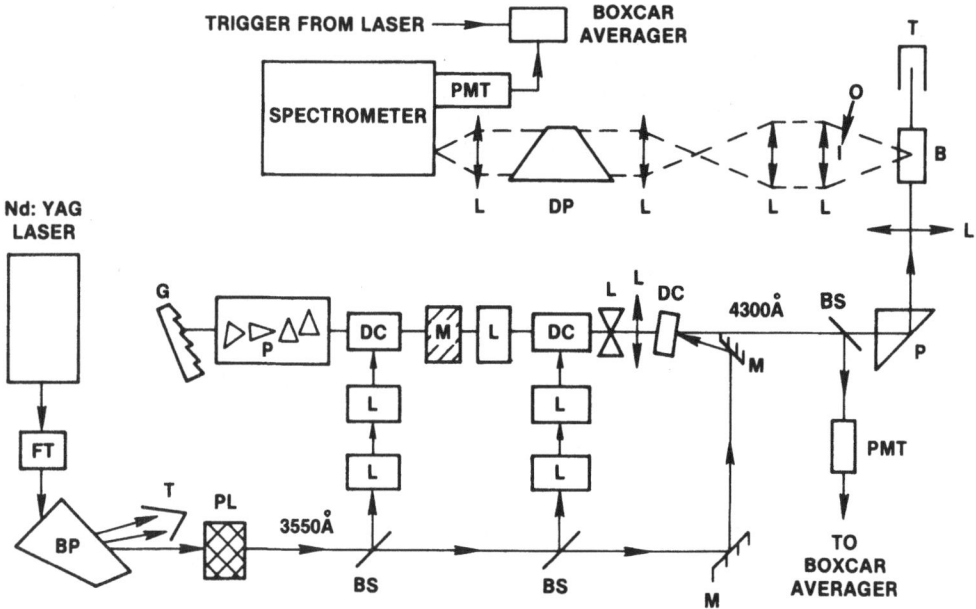

Fig. 3.   Schematic diagram of laser-induced saturated fluorescence
          apparatus: B, burner; BP, Brewster prism; BS, beam
          splitter; DC, dye cell; DP, dove prism; FT, frequency
          tripler; G, grating; L, lens; M, mirror; O, obscuration
          disk; P, prism; PL, half-wave plate; PMT, photomultiplier
          tube; T, light trap.

laser output with a flowing stilbene/methanol dye solution peaked
near 4270 Å which was particularly suitable for CH excitation. The
laser spectral width was about 0.3 Å. The burner in Figure 3 is a
modified oxy/acetylene welding torch capable of yielding (10 - 100)
ppm CH concentrations. The dye laser output is focussed into the
flame by a lens such that the focussed spot diameter is roughly 0.027
cm. Fluorescence is viewed at a right angle with respect to the
direction of laser propagation. The lenses and prism which constitute the light collecting optics rotate the horizontal laser/flame
interaction length onto the vertical spectrometer slit without magnification. The signal recording instrumentation in Figure 3 consists
of a photomultiplier tube and a boxcar averager.

     Data for CH fluorescence as a function of laser energy are
displayed in Figure 4. Fluorescence was observed very near 4308 Å,
which corresponds principally to laser-excited Q(9) emission at this
wavelength. The departure from linearity between 0.01 and 0.1 mJ is
indicative of saturation. The results of Figure 4 and Equation (8)
above were used to compute species concentration and excited-state
quenching. From a scan of the entire CH laser-induced spectrum, it

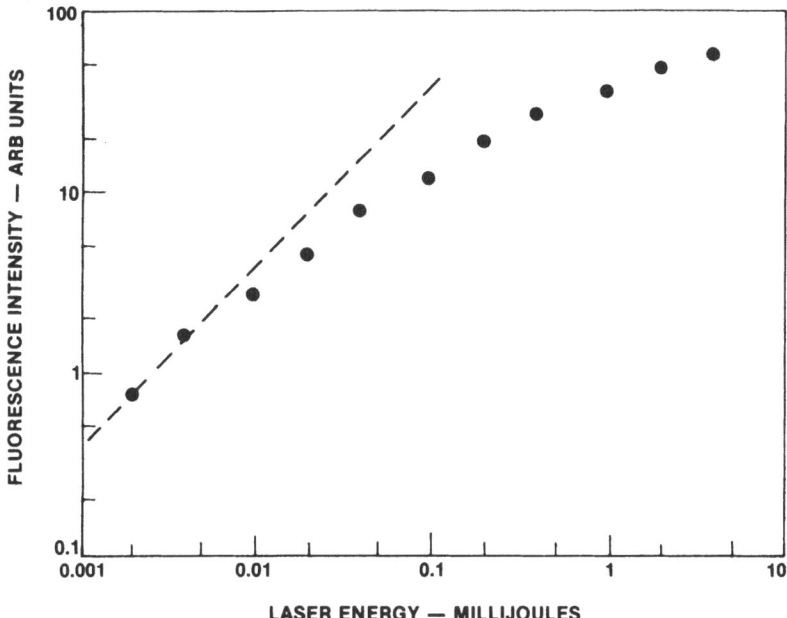

Fig. 4.   Laser-excited CH flame fluoresence versus laser energy for
the absorbing transition λ_{EXC} = 4262 Å [$^2\Pi(v''=0,K''=8)$ →
$^2\Delta(v'=0,K'=9)$].

was apparent from relative line intensities that rotational equili-
bration was not present. However, since rotational relaxation rates
were not available for CH, it was not possible to evaluate f precisely and hence it was taken as equal to the Boltzmann fraction f_B.
The results obtained for CH concentration and quenching are N_1° =
(103 ± 50)ppm (T = 2600°K) and Q_{21} = 8 x 10^{11} sec^{-1}. In order to
test the validity of the result for N_1°, it was determined in a
separate optical absorption measurement to be (57 ± 20) ppm[3]. This
result is in reasonable agreement with the result from fluorescence,
particularly in view of the uncertainty in f_B. In fact, the higher
fluorescence result may be due to rotational redistribution occurring
during the time of the pulse so that the fraction of molecules actually pumped, f exceeds f_B. The results for CH and several other
molecular species as well are compiled in Table 1.

Discussion

Saturated laser-induced fluorescence is a highly useful technique for species concentration measurement in combustion media.
There are certain factors, however, which may either invalidate the
technique or limit its accuracy. A brief discussion of several of
these factors follows below.

Table 1. Summary of Saturated Molecular Fluorescence Measurements

Molecule	Flame	N_1° (cm^{-3})	Q_{21} (sec^{-1})	Ref.
C_2	C_2H_2/O_2	4.5×10^{15}	12.1×10^{11}	8
CH	C_2H_2/O_2	3.3×10^{14}	8×10^{11}	3
CN	C_2H_2/N_2O	1.1×10^{15}	2×10^{13}	3
MgO	$Mg/C_2H_2/Air$	2.6×10^{11}	7×10^{10}	9
OH	CH_4/Air	4×10^{16}		10

The use of short laser pulses for fluorescence measurements is desirable in that they correspond to high laser power which is optimal for saturation. With short pulses, however, there is concern that a steady-state condition is achieved during the laser pulse; for the experiments reported on in [Reference 3], the pulse duration was 10 nsec. Using time-dependent solutions of Equations (1a) and (1b), it may be shown that upper and lower level populations are related by

$$N_2 = N_1 g_2/g_1 [1 - \exp\{ - (1 + g_2/g_1)B_{12}I_{L\nu}\tau\}], \hspace{1cm} (9)$$

where $B_{12}I_{L\nu} \gg (Q_{21} + A_{21})$ has been assumed. Steady-state is reached in a time $\tau \sim (B_{12}I_{L\nu})^{-1} \sim (Q_{21} + A_{21})^{-1}$ for saturation conditions. For CH in [Reference 3], for example, A_{21} $(A^2\Delta \rightarrow X^2\Pi) = 2 \times 10^6$ sec^{-1} and $Q_{21} = 8 \times 10^{11}$ sec^{-1} so that, τ (CH) $\sim 10^{-12}$ sec. In this case, the 10 nsec laser pulse is long enough to ensure a steady-state condition. It is important to note, however, that for applications where Q_{21} is considerably smaller than the value above, τ is correspondingly longer and steady-state may, in fact, not be realized.

From the preceding discussion concerning the steady-state assumption, it would appear that longer pulses, $O(1 \mu sec)$ characteristic of flashlamp-pumped dye lasers, may be desirable. This, however, is not true in that it may lead to other difficulties. Muller et al.,[11] have pointed out that laser-induced chemistry may occur on a time scale of 1 μsec, and that the chemistry would invalidate the approach outlined above since the reactions would deplete ground-state species. Muller et al., performed an analysis of Na and Li $(^2P_{1/2, 3/2} \rightarrow {}^2S_{1/2})$ fluorescence in atmospheric pressure $H_2/O_2/N_2$ flames. They concluded that reactions of the type, $Na(^2P_{1/2}) + H_2O \rightleftarrows NaOH + H$ and $Na(^2P_{1/2}) + H_2 \rightleftarrows NaH + H$, were significant on the time scale indicated, and recommended the use of nsec pulse length lasers to circumvent these difficulties.

The absence of numerical values for f in Equation (8) for various combustion species is another area of uncertainty. In order to evaluate f, rotational relaxation rates and their temperature dependence must be known for a broad range of rotational quantum numbers.

The difficulty is that there are almost no date of this type available. An exception to this is the recent work of Smith and
Crosley[12] for OH where the ratios of the quenching rate to the
total rotational energy transfer rate have been determined for a
range of rotational quantum numbers. More data of this type are
needed for other molecules.

In the treatment of saturated fluorescence above, it was assumed
that the interaction volume is uniformly illuminated. This implies a
rectangular spatial distribution for the laser intensity which is not
realistic. A Guassian distribution, on the other hand, is probably
not unreasonable. For such a case, however, if saturation occurs at
a point in space coincident with the peak of the distribution, it may
or may not occur in the wings of the distribution. Assuming the
laser spot size to be smaller than characteristic flame dimensions,
some molecules are saturated and others are not leading to errors in
measured species concentration. Pasternack et al.,[9] have analyzed
data for Na using rectangular, Gaussian and truncated Gaussian profiles and observed the influence of this variation on calculated
species concentrations. The rectangular profile gave a result for Na
more nearly equivalent to that determined from optical absorption
does not alter the fact that a satisfactory and consistent procedure
is needed to account for intensity distribution[13].

Despite the foregoing factors which can limit the accuracy of
saturated laser-induced fluorescence, the technique remains highly
valuable for combustion applications. It should be stressed that
measurements have been made successfully for extremely reactive
radical species in atmospheric pressure flames at temperatures of
0(2500°C) with reasonably good agreement between absorption and
fluorescence approaches. For example, absorption and fluorescence
measurements on OH are now in agreement within 20 percent[14].

B. PARTICULATE MIE SCATTERING

There is currently, a great deal of interest in discovering
means to control the quantity of particulate soot formed in practical
combustor emissions. The presence of soot within and in the exhaust
of a combustor is almost always undesirable. It is injurious to
human health, and soot causes an increased rate of heat transfer to
normally cool combustor surfaces resulting in their degradation. In
general, there is a need for diagnostic techniques which measure the
size and quantity of particulates in order to judge the efficacy of
various procedures implemented to control their formation. Tra-
ditional techniques, such as collection via impaction, electrostatic
or thermal precipitation and size measurement via optical or electron
microscopy, tend to be highly qualitative, perturbing and too slow.
There is a need for continuous, rapid, in-situ monitoring of particulate emissions. Optical techniques relying upon laser light scattering offer the potential to provide this capability.

There are three particulate properties of importance with regard
to signal intensities in a laser light scattering measurement. These
properties are size, shape and refractive index. The size of a
particulate is categorized qualitatively with reference to the wavelength of the light source. In Rayleigh scattering, the particulate
diameter D is smaller than a wavelength λ, i.e. $\pi D/\lambda < 1$, whereas for
Mie scattering the reverse is true, i.e. $\pi D/\lambda > 1$.[15] A significant
difference between the two regimes is that their angular patterns of
scattered intensity are very dissimilar[16]. The shape of particulates may vary from nearly spherical to very nonspherical. In the
Mie regime the mathematical solutions to light scattering for the
latter are very complex[17,18]. In some cases this tends to limit
their general usefulness. The refractive index value, particularly
the imaginary component, is indicative of the chemical nature of the
particulate. Fly ash, one product of coal combustion, consists
predominantly of SiO_2, Al_2O_3 and Fe_2O_3. Alumina (Al_2O_3), for
example, has an imaginary index component between 10^{-6} and 10^{-5} in
the visible making it highly transparent. On the other hand, carbonaceous particulates and hydrocarbon soots have imaginary components
in the 0.4 to 1.0 range making them highly absorbing.

The term Mie scattering is most often used loosely to include
both light scattering and extinction measurements. Here, a brief
outline will be given of the theory as it applies to size and number
density determination for a polydisperse size distribution of spherical scatterers[19]. Light extinction is given by

$$I_t = I_o \exp - [<Q(\hat{n},x)> NL], \tag{10}$$

where $<Q(\hat{n},x)>$ is an extinction cross-section, N is the total number
density of particulates, and L is the extinction pathlength. The
cross-section is a function of a complex refractive index $\hat{n}$, particulate diameter $D(x = \pi D/\lambda)$, and is averaged over a distribution function normalized to unity. Light scattering is given by

$$I_s = \frac{\lambda^2}{4\pi^2 r^2} <\phi(\hat{n}, x, \theta)_{s,p}> NV\Omega(I_o/A). \tag{11}$$

In Equation (11), $(I_s r^2)$ is the scattered power for a distance r,
from the scattering volume to the detector; $<\phi(\hat{n},x,\theta)_{s,p}>$ is a complicated function of $\hat{n}$, x, scattering angle θ, light polarization
direction "s" or "p", and is averaged over a distribution function; A
is the area of the incident beam of light; V and Ω are as in Equation
(3).

In a practical measurement application, one seeks to determine
the parameters of the size distribution, i.e. mean diameter, distribution width, and N. Other parameters in Equations (10) and (11) are
usually known, so that there are three unknown quantities. Clearly,
the particulate parameters cannot be determined fully from a single

extinction or scattering measurement and some combinations of these are required. Although there are applications where extinction measurements are highly useful and, for example, the wavelength dependence of extinction may be used to advantage for sizing, such discussions will be omitted here. In combustion media, the spatial distribution of particulates is very often not uniform and, hence, extinction measurements are invalid. Rather then, examples of size and number density determination will be given which relate to scattering measurements alone.

An intensity ratio useful for size determination is given by

$$I_p/I_s = \phi(\hat{n}, x_m, \sigma_0, \Theta)_p/\phi(\hat{n}, x_m, \sigma_0, \Theta)_s, \tag{12}$$

where D_m (= $x_m \lambda/\pi$) and σ_0 are the modal diameter (function peaks at D = D_m) and distribution width, respectively. Importantly, various instrumental factors evident in Equation (11) are not present in Equation (12). The quantity in Equation (12) is the ratio of scattered intensities at an angle θ, for two different polarization directions of the incident light; for fixed θ, I_p/I_s is a function of x_m and σ_0 alone since the dependences of ϕ_p and ϕ_s on the latter are dissimilar and $\hat{n}$ is assumed known. The ratio is given in Figure 5 as a function of D_m for a zeroth-order logarithmic distribution[15]. Each curve corresponds to a different width σ_0; it may be shown that I_p/I_s is most sensitive to D_m for $\theta = 90°$. From Figure 5 it is apparent that a single ratio measurement is not sufficient to determine both D_m and σ_0. In this connection the ratio of scattered intensities for two different angles and for "s" polarization given by

$$I_s(\Theta_1)/I_s(\Theta_2) = \phi(\hat{n}, x_m, \sigma_0, \Theta_1)_s/\phi(\hat{n}, x_m, \sigma_0, \Theta_2)_s \tag{13}$$

and shown in Figure 6 for $\theta_1 = 45°$ and $\theta_2 = 135°$ is useful. Since the dependences of the ratios in Figures 5 and 6 on D_m and σ_0 are dissimilar, the two ratios may be used to determine the latter parameters. This then, permits the evaluation of ϕ_s, which in turn may be used in Equation (11) to determine N. Finally, note that for a size monodispersion, i.e. $\sigma_0 = 0$, either ratio may be used to evaluate D.

There are two types of scattering measurements appropriate for size determination. One is the scattering from a cloud of particulates described above and referred to as global scattering, while the other is scattering from a single particulate. For the latter there is only one particulate in the optically sensitive scattering volume at any instant. This condition has a single, powerful advantage. By continuous monitoring of the scattered signals and electronic sorting of them in accordance with their size, it is possible to characterize the size distribution precisely. In contrast to global scattering, there is no need to assume a functional form for the distribution so

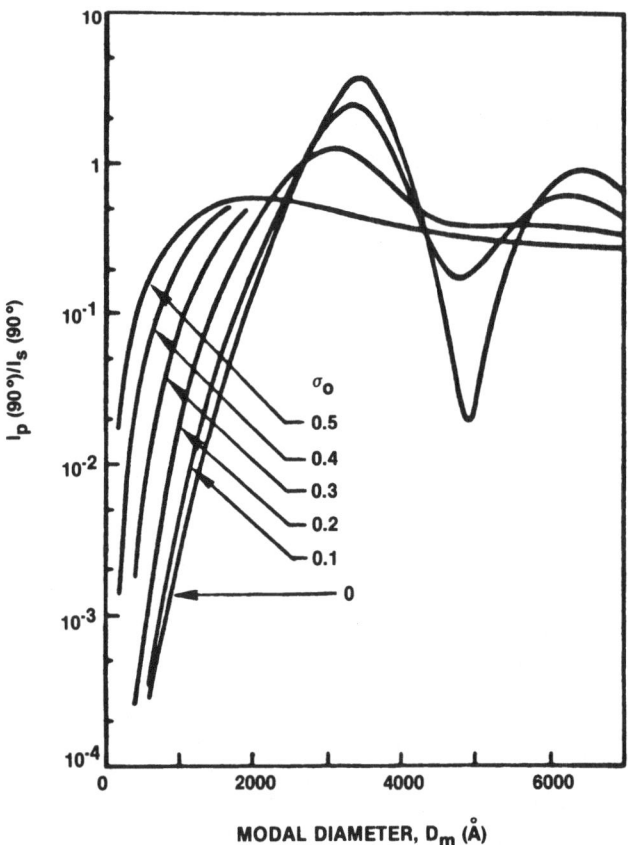

Fig. 5. Ratio of scattered-light intensities at 90° for incident
         light linearly polarized parallel (p) and perpendicular (s)
         to the scattering plane. The modal diameter, D_m, and σ_o are
         parameters in the zeroth-order logarithmic distribution
         function (ZOLD). Refractive index: $\hat{n} = 1.94-0.66\hat{\imath}$. Wave-
         length of light source: $\lambda = 5145$ Å. [Reprinted from Refer-
         ence 19 with the permission of The Combustion Institute].

that it may be, for example, bimodal or even more complex. There are
in fact, certain limitations to the single particle approach which
make it complementary rather than superior to global scattering. It
is extremely difficult to construct a sample volume such that, $V<10^{-7}$
cm^3. This limits single particle methods to media for which, $N<10^7$
cm^3. In addition the scattered intensity for small particulates
decreases with size proportional to D^6/λ^4; thus the intensities are
correspondingly small in contrast with global scattering where the
intensity is proportional to $NV(D^6/\lambda^4)$ and typically, $NV \sim (10^5-10^9)$.
The latter consideration limits single particle techniques to $D \geq$
$(0.1-0.2)\mu$. Global scattering, on the other hand, is applicable to

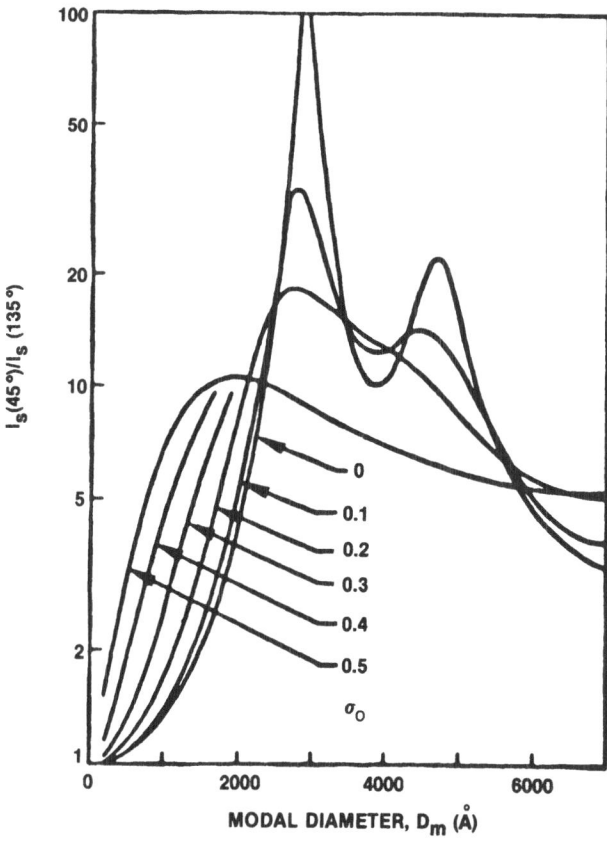

Fig. 6.   Ratio of scattered-light intensities at 45° and 135° for
incident light linearly polarized perpendicular to the plane
of scattering. Wavelength of light source: λ = 5145 Å.
[Reprinted from Reference 19 with the permission of The
Combustion Institute].

size measurement down to 50 Å and number densities up to 10^{12} cm^{-3},
thereby emphasizing the complementary nature of the two approaches.

Soot Particulates

Carbonaceous particulates appearing as combustion products are
characterized by: 1) irregular shape; 2) variable chemical composition; 3) broad size range; 4) inhomogeneous spatial distribution.
These factors make the application of light scattering techniques to
soot characterization more difficult than is the case for more well-
defined particulates. Soot particulates are chainlike in structure
as determined by their extraction from flames with collecting probes
and their subsequent examination via electron microscopy. Accord-

ingly they are neither spheres nor does their shape approximate that
of well-defined spheroid. Moreover, the exact particulate shape
varies in accordance with different stages of the combustion process.
In early stages of soot formation, the particulates are small and
approximately spheres or spheroids, whereas in later and final stages
they have the chainlike structure already mentioned. The chemical
composition of soot particulates may be classified in accordance with
their C/H ratio[20]. Again, this quantity depends on the stage of
the combustion process, with the C/H ratio being largest in the final
stage. It may be argued as well that fuel type influences the latter
ratio. As regards optical diagnostics, the variable C/H ratio results in an uncertainty in the refractive index. The size of soot
particulates may vary over a very broad range, and it is not neces-
sarily true that their distribution function is unimodal. Depending
on the application soot particulate diameters may range between 50 Å
and 10μ or even larger. In this regard, Mie scattering as outlined
in Equations (10) through (13) is limited to D < (0.1-0.2)μ. The
limitations on Mie scattering are apparent in Figures 5 and 6. For D_m
> (1000-2000) Å and σ_0 = (0.1-0.3), the ratios are multivalued func-
tions and as such are not, in principle, applicable to size determination. Fortunately in many cases the particulates do not exceed the
latter size. Finally, as with molecular/radical species, soot particulates are often dispersed inhomogeneously in combustion media,
and flame incandescence is a source of signal interference.

The approaches taken to deal with the preceding difficulties are
briefly as follows. The exact theoretical treatment of light scattering by soot particulates is prohibitively complex in view of their
very irregular shape. Treating them approximately as spheroids is
not desirable, as it turns out, since this overly increases the
number of parameters required to characterize the particulates fully.
Rather the particulates are treated as spheres, and the measured
diameters are interpreted as approximate particulate sizes. Although
this may seem to be an inaccurate and dubious procedure, it in fact
is not. For small particulates such that x (= $\pi D/\lambda$)<1, a condition
often satisfied in flames, light scattering is rather insensitive to
shape. In addition, for particulates whose orientation in space is
random due to a tumbling motion and whose size is characterized by a
broad distribution function, the difference between spherical theory
and experimental results are small[21].

There are two options to select from in dealing with refractive
index uncertainties. First, it is possible in principle to treat the
real and imaginary parts of the index as unknown parameters, and
attempt to evaluate them from scattering measurements supplemental to
and independent of those in Equations (12) and (13). For well defined particles, like certain polymeric spheres, this is a viable
procedure. For soot particulates, however, it is not since the
number of variables becomes excessive. The second option is to take
the index value from independent determinations such as optical

reflectance measurements of ñ. This in fact is a reasonable, ap-proximate procedure whose validity is enhanced by selecting angular regions of scattering which are relatively insensitive to index variation[19].

In cases where the particulate size distribution is very broad and there are significant numbers of large particulates, it is neces-sary to supplement Mie scattering with other techniques such as laser doppler velocimetry (LDV)[22]. LDV applied to particulate sizing is a single particle technique based on interferometric principles. In LDV, interference between the wave fronts of two laser beams forms a series of fringes parallel to the plane of their interaction. As a particulate crosses the fringe volume, a modulated scattering signal is produced which may be related to particulate size. LDV is applic-able to particulates present at modest number densities and having $1 < D < 10\mu$.

The spatial inhomogeneities associated with soot formation do not constitute a fundamental difficulty. Extinction measurements should be avoided, and a small optical sample volume should be de-fined as mentioned above in connection with saturated fluorescence. Normal soot incandescence emission is a source of signal interference in Mie scattering in flames. This interference is overcome by using a visible cw laser as a light source and narrow-band optical filters centered at the laser wavelength and positioned in front of light scattering detectors so as to discriminate against incandescence. Since cw lasers are used with output power $\sim$ 1 Watt, laser modulated particulate incandenscence is not a signal interference source in Mie scattering.

Particulate Measurements

In Figure 7, an apparatus is shown which was used successfully to characterize the size and number density of soot particulates formed during the pyrolysis of acetylene[19]. The pyrolysis cell in Figure 7 is a cylindrical quartz tube raised to a temperature of $\sim$ 1000°C by an electrical heating tape. A C_2H_2/He mixture flows through the tube. The function of the He is to raise the flash-point temperature of the mixture. The cell has five arms for entrance and exit of the laser beam and for scattering measurements at 45°, 90° and 135°. The light source is a cw ar$^+$ laser operating at 5145 Å with a 1 Watt output power. The laser output is linearly polarized to a very high degree, and the polarization rotator in Figure 7 permits continuous rotation of the polarization direction. The "s" and "p" directions referred to above correspond to laser polarization perpendicular and parallel to the scattering plane, respectively, where this plane is formed by the intersection of the laser beam axis and the line between the sample volume and one of the detectors. The scattering arms have apertures which set the light collection solid

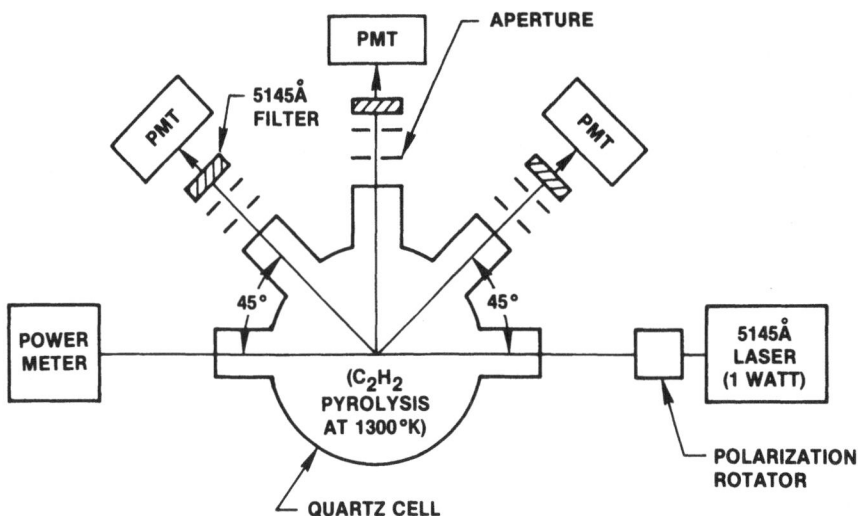

Fig. 7. Schematic diagram of Mie scattering apparatus. [Reprinted
from Reference 19 with the permission of The Combustion
Institute].

angle and sample volume at $\Omega = 10^{-5}$ and $V = 0.014$ cm^3, respectively.
Attached to each of the photomultiplier tubes (PMT) are polarizers
whose axes are parallel to the laser polarization direction.

The soot particulate parameters D_m and σ_0 for acetylene pyrolysis were obtained from intensity ratios given in Equations (12) and
(13). The number density N was determined from Equation (11) after
evaluating ϕ ($\hat{n} = 1.94-0.66\hat{i}$; $\Theta = 45°$) and inserting numerical
values for V, Ω, A and the absolute sensitivity of the photomultiplier. The latter apparatus parameters were determined accurately as
follows. The pyrolysis cell was replaced by a liquid suspension of
polymeric spheres whose size, concentration and refractive index were
precisely known. The absolute scattering intensity at 45° was used
then in Equation (11) to obtain the product, $(V\Omega/A)\varepsilon$, where ε is the
photomultiplier sensitivity[23,24].

The soot particulate mean diameter and mass fraction are shown
in Figure 8 as a function of acetylene flow rate. For the ZOLD
function, the mean diameter $\bar{D}$ is given by $\bar{D} = D_m \exp (1.5\sigma_0^2)$. The
distribution width for different flow rates was consistent with $\sigma_0 =$
$(0.4{+0.0 \atop -0.2})$. The mass fraction in Figure 8 is $\rho\pi(\bar{D})^3 N/6$, where $\rho = 2.2$
g/cm^3 is the particulate density. Soot particulates tend to agglom-
erate and increase their size due to collisions. The growth in size
in Figure 8 is due to increasing collision frequency resulting from
increased fuel flow rate. Using the apparatus in Figure 7, data have
been obtained for the effect of trace amounts of O$_2$ and NO on carbon

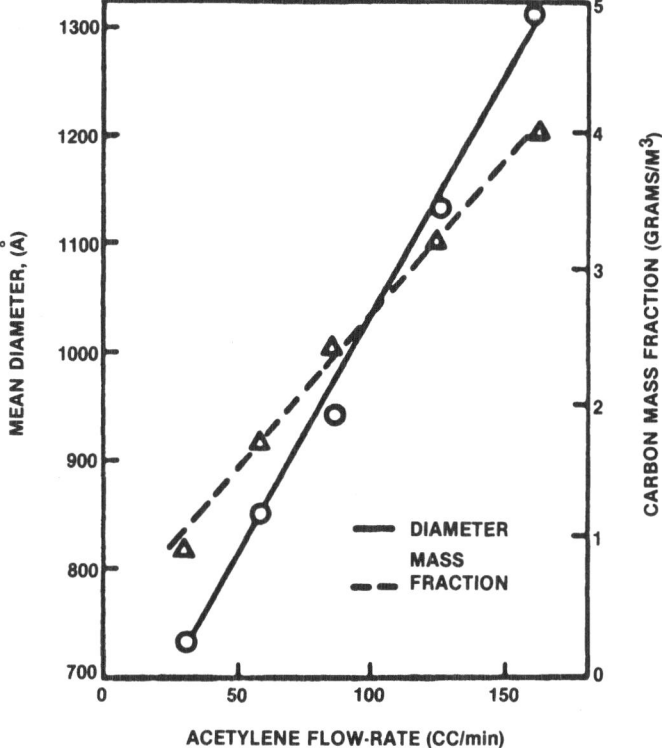

Fig. 8.   Carbon mass fraction and particulate mean diameter versus
          acetylene flow rate [Reprinted from Reference 19 with the
          permission of The Combustion Institute].

formation for a fixed acetylene flow rate. The latter species act to
reduce particulate diameter, number density and mass fraction due to
rapid particulate burn-out via, for example, $C + (1/2)O_2 \rightarrow CO + 26.4$
kcal/mol.

     In addition to fuel pyrolysis applications, Mie scattering has
been used to characterize soot particulates in flames. D'Alessio et
al.,[25] have made measurements in atmospheric, premixed flat CH_4/O_2
flames using scatter and extinction. In the very special case of
flat flames, extinction measurements are possible since the medium is
radially homogeneous. The scattering measurements were made at 20
and 160°, and axial (vertical) profiles of D and N were determined
along with an approximate evaluation of σ_0. In recent experiments
done by the author at UTRC, Mie scattering has been used to determine
D and N for soot particulates in a laminar $C_3H_8/O_2/N_2$ diffusion flame
with and without metallic additives present. The principal effect of
the additives was to alter $\bar{D}$ and N. For certain combinations of
additive concentration and fuel/air ratio, the soot volume fraction

was reduced significantly below that occurring in the unseeded flame, which is an important result with respect to controlling soot formation. The particulate parameters were determined from angular dissymmetry measurements at 45/135°. Results for LiCl and KCl addition are shown in Figure 9. It may be shown that the more pronounced effect of KCl on D is due to the lower ionization potential of K relative to Li, and the resulting higher concentration of K^+ which act to retard particulate growth.

The measurements discussed above have been all global in nature. Hirleman and Wittig[26] have applied single particle techniques to characterize particulates emitted in automotive exhaust gases. The scattered light intensities were measured in the near forward direction at 3°, 6° and 12°. It may be shown that the ratios 6°/3° and 12°/6° are effective, taken in tandem, for the precise and unambiguous determination of size in the range, $0.4 < D < 3.0\mu$. It may be shown as well that near forward scattering and ratioing of the intensities minimizes the dependence of the results on refractive index uncertainties. Size distributions were obtained for $V \sim 3 \times 10^{-6}$ cm^3, which sets an upper limit on concentration given by $N_{max} \sim 3 \times 10^5$ cm^{-3}. Distributions from light scattering were only in fair agreement with those obtained from probe sampling, but the two measurements were not made in parallel on a single engine.

Discussion

In the preceding outline of Mie scattering theory and applications, various approximate procedures were discussed, which were made necessary by the complex physical nature of soot particulates. Clearly it is highly desirable to assess the validity of these approximations in order to judge the accuracy of measurements. In the case of soot, one way to approach this is to compare light scattering results with other measurements; several such comparisons are given immediately below.

Bockhorn et al.,[27] have made soot concentration and particulate size measurements via both light scattering and extraction probe measurements in atmospheric C_3H_8/O_2, $C_3H_8/O_2/N_2/H_2$, and $C_3H_8/O_2/N_2/NH_3$ flames. The flames to which H_2 and NH_3 were added had lower soot concentrations characterized by a smaller mean diameter and a reduced distribution width. The objective of this work was to determine the mechanisms responsible for soot nucleation, growth and coagulation in flames. The effects of additives such as H_2 and NH_3 can be helpful in understanding these mechanisms. The comparison by Bockhorn et al., of light scattering and probe results are of particular interest. For the C_3H_8/O_2 flame, the soot concentration was relatively high, i.e. $\sim 3 \times 10^{-6} g/cm^3$. In this case, the sizes from light scattering were smaller than those from probe measurements. The mean diameter from light scattering was 1200 Å, whereas from probe meas-

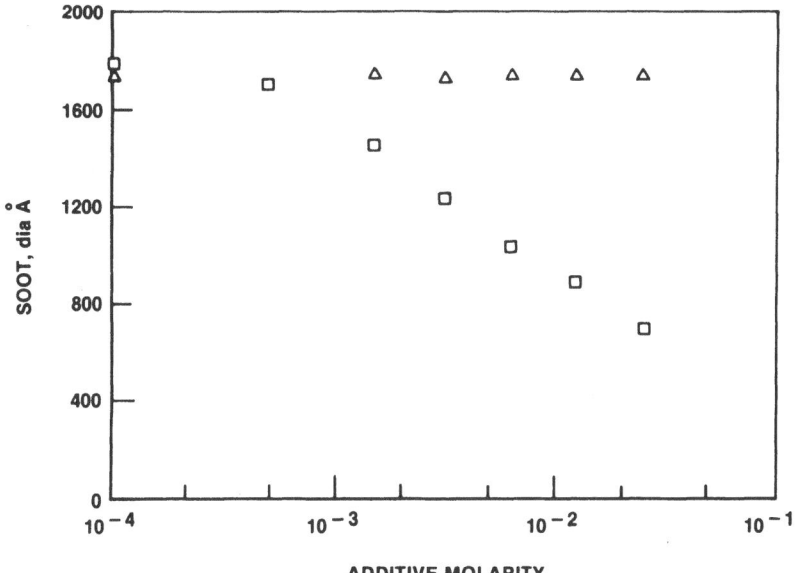

Fig. 9.   Dependence of soot particulate diameter on additive molarity
          for $C_3H_8/O_2/N_2$ diffusion flame at a temperature, T = 1973°K.
          Additive: Δ, LiCl; ▢; KCl.

urements it was 1600 Å. For the flames to which H_2 and NH_3 were
added, the soot concentrations were lower, i.e. ~2 x 10^{-6}g/cm^3, and
the agreement between the two approaches was excellent as regards
size determination. The results of this work generally validate a
light scattering measurement approach to particulate characterization. For combustion applications, even the 400 Å discrepancy
above is not regarded as very serious.

Hack et al.,[28] have compared optical and probe techniques for
soot formation in a swirl-stabilized combustor. The combustor simul-
ates important, practical conditions in that there occur complex
aerodynamic effects characterized by turbulence and reverse flow.
Here as in [Reference 27] quite reasonable agreement was obtained
between probe and optical methods for both size and number density.
There was, however, an important caveat; namely positioning of the
probe within the recirculation zone of the combustor can lead to
perturbations resulting in very significant changes in local particulate number density which then invalidates the comparison of techniques. In fact, this reemphasizes the importance of optical sampling approaches, in that they are non-perturbing.

In evaluating the accuracy of Mie scattering measurements, yet
another approach is to compare them with other, variant optical
techniques. Such an approach was used by Chang and Penner[29] who

compared angular dissymmetry and diffusion broadening spectroscopy
measurements of particulate size. Angular dissymmetry Mie measurements were discussed above. In diffusion broadening spectroscopy,
the homodyne beating of the discrete frequencies in a light scattering spectrum is analyzed and particulate diameter is inferred from
a Lorentzian-shaped response function. The technique is global, and
is applicable to sizing of a very small particulates. In [Reference
29] excellent agreement was obtained between the two optical approaches for particulate diameters in the range, $150 < D < 600$ Å, present
at various heights in a premixed CH_4/O_2 flame.

The preceding discussion illustrates that Mie scattering is a
very viable approach to particulate characterization. Even for
ill-defined soot particulates, sizing may be carried out with reasonable accuracy and without medium perturbations. As such the technique is far superior to older, classical methods.

REFERENCES

1. A. C. Eckbreth, P. A. Bonczyk, and J. F. Verdieck, Combustion
    Diagnostics by Laser Raman and Fluorescence Techniques,
    Prog.Energy Combust.Sci., 5:253 (1979).
2. A. P. Baronavski and J. R. McDonald, Application of Saturation
    Spectroscopy to the Measurement of $C_2,^3\Pi_u$ Concentrations in
    Oxy-Acetylene Flames, App.Opt., 16:1897 (1977).
3. J. F. Verdieck and P. A. Bonczyk, Laser-Induced Saturated
    Fluorescence Investigations of CH, CN and NO in Flames, in:
    "Eighteenth Symposium (International) on Combustion", The
    Combustion Institute, Pittsburgh (1981).
4. J. O. Berg and W. L. Shackleford, Rotational Redistribution
    Effect on Saturated Laser-Induced Fluorescence, Appl.Opt.,
    18:2093 (1979).
5. R. P. Lucht, D. W. Sweeney, and N. M. Laurendeau, Balanced Cross
    Rate Model for Saturated Molecular Fluorescence in Flames
    Using a Nanosecond Pulse Length Laser, Appl.Opt., 19:3295
    (1980).
6. A. C. Eckbreth and J. W. Davis, Spatial Resolution Enhancement
    in Coaxial Light Scattering Geometries, Appl.Opt., 16:804
    (1977).
7. D. Stepowski and M. J. Cottereau, Direct Measurement of OH Local
    Concentration in a Flame from the Fluorescence Induced by a
    Single Laser Pulse, Appl.Opt., 18:354 (1979).
8. A. P. Baronavski and J. R. McDonald, Measurement of C_2 Concen-
    trations in an Oxygen-Acetylene Flame: An Application of
    Saturation Spectroscopy, J.Chem.Phys., 66:3300 (1977).
9. L. Pasternack, A. P. Baronavski, and J. R. McDonald, Application
    of Saturation Spectroscopy for Measurement of Atomic Na and
    MgO in Acetylene Flames, J.Chem.Phys., 69:4830 (1978).

10. R. P. Lucht, D. W. Sweeney, and N. M. Laurendeau, Saturated-
     Fluorescence Measurements of the Hydroxyl Radical, in: "Laser
     Probes for Combustion Chemistry", D. R. Crosley, ed.,
     American Chemical Society, Washington (1980).

11. C. H. Muller III, K. Schofield, and M. Steinberg, Laser Induced
     Flame Chemistry of $Li(2^2P_{1/2,3/2})$ and $Na(3^2P_{1/2,3/2})$. Impli-
     cations for other Saturated Mode Measurement, J.Chem.Phys.,
     72:6620 (1980).

12. G. P. Smith and D. R. Crosley, Quantitative Laser-Induced Fluor-
     escence in OH: Transition Probabilities and the Influence of
     Energy Transfer, in: "Eighteenth Symposium (International) on
     Combustion", The Combustion Institute, Pittsberg (1981).

13. J. W. Daily, Saturation of Fluorescence in Flames with a
     Gaussain Laser Beam, Appl.Opt., 17:225 (1978).

14. R. P. Lucht (Private Communication).

15. M. Kerker, in: "Scattering of Light and Other Electromagnetic
     Radiation", Academic Press, New York (1969).

16. M. Born and E. Wolf, in: "Principles of Optics", 4th ed.,
     Pergamon, New York (1970).

17. S. Asano and M. Sato, Light Scattering by Randomly Oriented
     Spheroidal Particles, Appl.Opt., 19:962 (1980).

18. D. W. Schuerman, R. T. Wang, B. A. S. Gustafson, and R. W.
     Schaefer, Systematic Studies of Light Scattering. 1: Particle
     Shape, Appl.Opt., 20:4039 (1981); Appl.Opt., 21:369 (1982).

19. P. A. Bonczyk, Measurement of Particulate Size by In Situ Laser-
     Optical Methods: A Critical Evaluation Applied to Fuel-
     Pyrolyzed Carbon, Combust.Flame, 35:191 (1979).

20. R. C. Millikan, Optical Properties of Soot, J.Opt.Soc.Am.,
     51:698 (1961).

21. R. G. Pinnick, D. E. Carroll, and D. J. Hofmann, Polarized Light
     Scattered from Monodisperse Randomly Oriented Nonspherical
     Aerosol Particles: Measurements, Appl.Opt., 15:384 (1976).

22. W. M. Farmer, Measurement of Particle Size, Number Density, and
     Velocity Using a Laser Interferometer, Appl.Opt., 11:2603
     (1972).

23. W. Heller and R. Tabibian, Experimental Investigations on the
     Light Scattering of Colloidal Spheres. IV. Scattering Ratio,
     J.Phys.Chem., 66:2059 (1962).

24. J. P. Kratohvil and T. P. Wallace, Calibration of Light Scat-
     tering Photometers. VII. Calibration by Means of Colloidal
     Dispersions of Mie Scatterers, J.Phys.D.:Appl.Phys.(Great
     Britain) 3:221 (1970).

25. A. D'Alessio, A. Di Lorenzo, A. F. Sarofim, F. Beretta, S. Masi,
     and C. Venitozzi, Soot Formation in Methane-Oxygen Flames,
     in: "Fifteenth Symposium (International) on Combustion",
     The Combustion Institute, Pittsburg (1974).

26. E. D. Hirleman Jr., and S. L. K. Wittig, In Situ Optical
     Measurement of Automobile Exhaust Gas Particulate Size
     Distributions: Regular Fuel and Methanol Mixtures, in:
     "Sixteenth Symposium (International) on Combustion",
     Pittsburg (1976).

27. H. Bockhorn, F. Fetting, U. Meyer, R. Reck, and G. Wannemacher,
     Measurement of the Soot Concentration and Soot Particle Sizes
     in Propane Oxygen Flames, in: "Eighteenth Symposium (Inter-
     national) on Combustion", Pittsburgh (1981).
28. R. L. Hack, G. S. Samuelsen, C. C. Poon, and W. D. Bachalo,
     An Exploratory Study of Soot Sample Integrity and Probe
     Perturbation in a Swirl-Stabilized Combustor, ASME J.Eng.
     Power, 103:759 (1981).
29. P. H. P. Chang and S. S. Penner, Particle-Size Measurements in
     Flames Using Light Scattering; Comparison with Diffusion-
     Broadening Spectroscopy, J.Quant.Spectrosc.Radiat.Transfer,
     25:105 (1981).

ANALYTICAL AND DIAGNOSTIC APPLICATIONS OF LASER

INDUCED FLUORESCENCE IN FLAMES AND PLASMAS

N. Omenetto

Joint Research Center
Chemistry Division
Ispra (Varese), Italy

1. INTRODUCTION

The use of tunable lasers has had a significant, if not unique, impact in atomic and molecular fluorescence spectroscopy because of the analytical potential of this source and of the possibility of non-intrusive optical probing of several physical parameters of fundamental interest in laboratory flames and industrial devices for a better understanding of their combustion kinetics.

The lecture is divided in two parts. Firstly, the analytical features of laser excited atomic fluorescence in atmospheric pressure flames and/or other atom reservoirs will be discussed with particular emphasis on analytical figures of merit such as limits of detection, freedom from interferences and range of applicability. Secondly, several methods based on the measurement of selected fluorescence transitions in the saturation regime will be shown to be potentially useful for obtaining spatially and temporally resolved information on parameters such as the flame temperature, the total number densities of species and the quantum efficiency of the transition.

2. ANALYTICAL CONSIDERATIONS[1-6]

The most attractive characteristic of the dye lasers in view of their application to analytical atomic fluoroescence spectroscopy were certainly the tunability with respect to wavelength and the power available within the spectral absorption bandwidth. This source was therefore considered to be ideal (although costly) replacement of the conventional excitation sources used in fluorescence work, which were operated in both pulsed (xenon lamps, hollow cathode lamps) and

continuous wave conditions (electrodeless discharge lamps, Eimac
Xenon arcs). In addition, the attainment of optical saturation of
the fluorescence signal resulted in a number of peculiar theoretical
features that are now critically discussed in the following considerations.

2.1. Independence of the Fluorescence Signal upon Source Fluctuations and Atomizer Quantum Efficiency

This outcome can be theoretically seen from the general relationship[7] between the ratio of the atomic populations (see Figure
1a) derived with the assumption of steady state conditions and when
the excitation intensity grows very high, i.e.

$$\frac{n_2 g_1}{n_1 g_2} = \frac{6.6 \times 10^3 \lambda_{12}^5 Y_{21} E_{\lambda_{12}}}{1 + 6.6 \times 10^3 \lambda_{12}^5 Y_{21} E_{\lambda_{12}}} \tag{1}$$

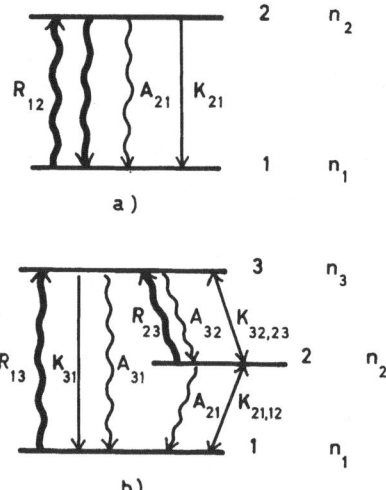

Fig. 1.   Simplified energy scheme for a 2-level (1a) and a 3-level
(1b) atomic system. n_1, n_2, and n_3 are the population
densities (cm^{-3}) of the levels whose statistical weights are
g_1, g_2 and g_3. The radiative rate coefficients (sec^{-1}) are
indicated as R_{13} for the laser excitation and as A_{21}, A_{31}
and A_{32} for the fluorescence. The collisional rate coefficients, both quenching and mixing, (sec^{-1}) are indicated
as k_{12}, k_{21}, k_{31}, k_{32} and k_{23}. $n_T = n_1 + n_2$ or $n_T = n_1 + n_2$
+ n_3 represents the total number density (cm^{-3}).

In this Equation, the following quantities are defined:

$E_{\lambda_{12}} \equiv \dfrac{\Phi_\ell}{\delta\lambda_\ell S_\ell} =$ spectral irradiance of the laser, considered as a continuum source with respect to the atomic absorption profile. Here Φ_ℓ (W) is the laser power, $\delta\lambda_\ell$ (cm) is the laser spectral bandwidth and S_ℓ (cm^2) is the laser cross section;

$Y_{21} = \left(\dfrac{A_{21}}{A_{21} + k_{21}}\right) =$ quantum efficiency of the transition;

A_{21}, k_{21} = Einstein coefficient of spontaneous emission and collisional quenching coefficient, sec^{-1}, respectively.

Therefore, one can see that the population of the excited level depends upon the <u>product</u> of the laser irradiance and the quantum efficiency. If this product grows much higher than unity, then $n_2 g_1 = n_1 g_2$ and if the statistical weights are the same, $n_2 = (n_T/2)$ when n_T is the total number of atoms in the (closed) two level system considered. As a consequence, the fluorescence signal (proportional to n_2) reaches a plateau and becomes independent on the laser irradiance.

In order to reach saturation, the laser induced radiative rate, R_{12}, must be greater than the quenching rate, k_{21}. Classical considerations[1,5] result in the following relationships

$$R_{12} = \frac{g_2}{g_1} A_{21} \left(\frac{\lambda_{12}^5}{8\pi hc^2} \right) E_{\lambda_{12}} \tag{2}$$

and

$$k_{21} = \sum_j n_j \sigma_j \bar{v}_j \tag{3}$$

where n_j is the density of the quenching species, σ_j the cross section for the quenching process and $\bar{v}_j$ the mean relative velocity of the collision partners. Typical values for flames at atmospheric pressure are $n_j \simeq 10^{18}$ cm^{-3}, $v_j \simeq 10^5$ cm sec^{-1} and $\sigma_j \simeq 10^{-15}$ cm^2, which give $k_{21} \simeq 10^8$ sec^{-1}. Therefore, one can calculate that the excitation rate resulting from the use of a Xenon arc ($\sim 10^4$ W cm^{-2} nm^{-1}) at $\lambda_{12} = 400$ nm and for a strong transition ($A_{21} \sim 10^8$ sec^{-1}) is approximately 10^4 s^{-1} while that of a pulsed laser having 10 kW cm^{-2} over a bandwidth of 0.03 nm is approximately 10^9 sec^{-1}. However, it is worth pointing out here that R_{21} varies with wavelength: for the same laser power and spectral bandwidth, R_{12} turns out to be

5.4×10^8 sec^{-1} at 300 nm and 7×10^7 sec^{-1} at 200 nm. Therefore, much higher laser powers are needed to saturate optical transitions down in the ultraviolet.

On the other hand, some further considerations are necessary and should not be overlooked. The foregoing theoretical predictions on the independence of the fluorescence signal upon the fluctuations of the laser source and the quantum efficiency of the atomizer hold for the rather unrealistic approximation of a unit step excitation pulse for the laser irradiance (i.e. $E_\lambda = 0$ for $t < 0$ and E_λ^o for $t > 0$) and for a measuring system that would faithfully reproduce the temporal behavior of n_2 (t). When boxcar integrators are used (which is quite common in pulsed[8] that, even for a rectangular laser pulse of duration Δt_ℓ, the signal will depend again upon the source fluctuations and the quantum efficiency whenever the gate width of the measuring device is larger than the width of the fluorescence pulse. Indeed (see Figure 2) one can easily derive for the excited state population the following expression[8]

$$\int_0^\infty n_2(t)dt = \int_0^{\Delta t_\ell} n_2(t)dt + \int_{\Delta t_\ell}^\infty n_2(t)dt =$$

$$= \left(\frac{E_\lambda^o n_\tau}{gE_\lambda^o+A}\right)\left\{\Delta t_\ell + \left[1-e^{-(gE_\lambda^o+A)\Delta t_\ell}\right]\left[\frac{1}{A} - \frac{1}{gE_\lambda^o+A}\right]\right\} \tag{4}$$

where

$$A \equiv A_{21}+k_{21} \text{ and } g \equiv \left(1 + \frac{g_1}{g_2}\right).$$

From this expression, one can see that even in the saturation limit (i.e. when E_λ^o grows to infinity) n_2 (t), and therefore the fluorescence signal, is given by

$$(n_2)_{E_\lambda^o \to \infty} = \left(\frac{n_\tau}{1+\frac{g_1}{g_2}}\right)\left(\Delta t_\ell + \frac{1}{A_{21}+k_{21}}\right) \tag{5}$$

and thus is still dependent upon the quantum efficiency of the transition via the second term in parenthesis. It would then be advisable to provide the possibility of changing the measuring gate width, ideally along the entire fluorescence waveform, provided that the signal-to-noise ratio is still acceptable.

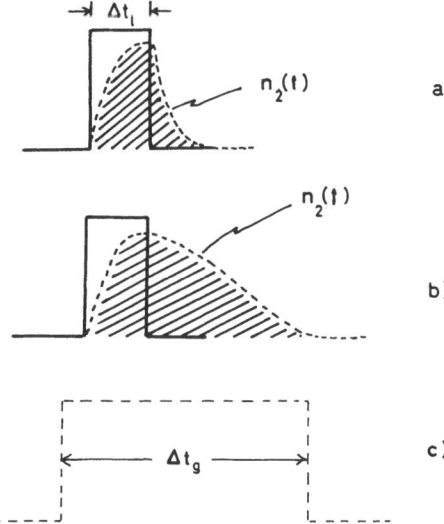

Fig. 2. Schematic representation of the measurement performed by a
boxcar integrator whose gate width, Δt_g, is larger than the
fluorescence pulse, proportional to $n_2(t)$, excited by a
rectangular laser pulse of duration Δt_ℓ. a) high quenching
flame; b) low quenching flame; c) gate width. It can be
noticed that, assuming saturation, the time response of the
signal within the laser pulse is the same, irrespective of
quenching.

2.2. Improved Linear Dynamic Range of the Analytical Calibration Curves

It is well known[1] that the fluorescence curves of growth are
characterized at high concentrations by a zero slope (in case of a
continuum excitation source) or a negative slope (in case of a line
excitation source). This behavior is essentially due to the self-
absorption factor (radiation trapping) which decreases the fluorescence radiation on its way towards the detection system. It should
be stressed here that the atomizer is assumed to be completely illuminated by the source while the entire fluorescence volume is seen
by the detector: in this case, geometrical artifacts such as pre-
filter and postfilter effects do not enter into consideration. When
saturation is reached the absorption coefficient, which is proportional to the difference between the atomic populations of the
levels, approaches zero. As a consequence, the medium becomes transparent and no self-absorption can occur. The improved linearity of
the analytical curves has been shown experimentally[9]. However, one
should realize that if the laser beam is focussed at the center of
the atomizer, the presence of self-reversal, which will affect the
shape of the calibration curves, must be considered.

2.3. Signal-to-Noise Improvement for Background Noise Limited Systems

Because of the gated operation of the detection system, the background is only measured during the opening of the gate at a frequency dictated by the repetition rate of the laser. In this case, the low duty cycle characteristic of the most pulsed lasers used for analytical applications should result in a significant improvement in the signal-to-noise ratio as compared to a cw operated source and a cw detection system, but only if the average spectral irradiance of the pulse source equals that of the cw source. Moreover, the theoretical improvement is predicted by assuming that, during the pulse, the fluorescence follows linearly the source power. Because of saturation, the increase in the fluorescence is (much) less and then the comparison will always be in favor of the cw system. On the other hand, the wavelength coverage (fundamental and frequency doubled) provided by pulsed lasers is still larger than that achieved with cw lasers.

Extensive considerations related to this topic can be found in the literature[1,3,5,10].

2.4. Analytical Sensitivity and Detection Limits

If the theoretical considerations made previously (Paragraph 2.1) hold, the maximum fluorescence signal will be measured when saturation is achieved, since for equal statistical weights, the population of both levels will be halved. Extensive theoretical calculations have been made aimed at giving order of magnitude estimates of the minimum detectable concentration of an atomic species in conventional analytical atomizers[2,5,11]. The calculations are based on signal-to-noise ratio considerations and the detection limits are referred to that concentrations level (usually in aqueous solution) which would give a signal to root-mean-square noise ratio of 3. As clearly pointed out by Alkemade[2] one should distinguish between "extrinsic" and "intrinsic" limit of detection, the difference being given by the fact that in the last case the main limiting noise is not due to the background or to the detector but to the inherent statistical fluctuation in the number of atoms present in the volume probed by the laser. Indeed, it has been reported in the literature[2,5,11] that such number can be extremely small so that single atom detection seems to be within the reach of the fluorescence technique. The reader is warned, however, that this spectacular sensitivity has been achieved in low pressure vapor cells and/or specially designed quartz containers where the atoms are surrounded by an inert atmosphere. In conventional atomizers at atmospheric pressure, the sensitivity is expected to be much poorer. In addition, the low number of atoms in the probe volume usually means that many more atoms ($\sim 10^3$x) are needed in the atomizer volume, which is finally what matters for analytical purposes.

It is worth remembering that the calculated limit of detection
is given, in number of atoms per unit volume, and therefore needs to
be related to the concentration (usually in /µg/mL or ppm) of the
element in the solution aspirated into the flame or plasma or introduced into the electrothermal atomizer. The conversion factor calculated for typical flame conditions is approximately 10^{11} atoms
cm^{-3}/ppm.

Table 1 collects the experimental detection limits obtained so
far from flames and electrothermal atomizers such as open carbon
cups. As one easily sees, even if work can still be done to improve
flame detection limits, the flame should be abandoned in favor of
electrothermal atomizers. This is even more substantiated by the
data collected in Table 2 which are especially important when only
microliter quantities of sample are available. Considering the
actual atomizer volume probed by the laser, one can calculate[12]
that the detection limit for lead corresponds to approximately 250
atoms/cm^3.

As repeatedly pointed out in the literature, Mie scattering is
the major problem in laser excited fluorescence. Such scattering
should not be confused with the (unavoidable) spurious reflections
(from the optical components) of the laser beam into the fluorescence
monochromator that can be minimized by proper and careful positioning
of light baffles, traps and possibly polarizers, but is rather due to
the presence of unvaporized particles created in the atomizer whenever the element sought is present in a large excess (1:10,000) of
matrix. The most effective way of reducing the scattering effect is
by the use of non resonance fluorescence transitions (see Figure 3).
Indeed, the best detection limits reported in Table 1 and 2, especially with electrothermal atomization, have been obtained by
measuring the fluorescence at a wavelength different from the excitation one.

Other ways have been proposed to correct for scattering[2,13] by
taking advantage of the different frequency content of the saturated
fluorescence and scattering signals. Time resolution is practically
useless in strongly quenching atomizers because of the reduced lifetime of the excited fluorescent level, and polarizers cannot be very
effective for Mie scatter as it would be in the case of Rayleigh
scatter. In conclusion, if only resonance fluorescence transitions
are available (Zn, Cd) the scattering will seriously detract from the
attractiveness of the technique. These considerations indicate the
importance of improving the analytical performances of the atomizer
and work should proceed in this direction.

2.5. Other Analytical Considerations

Spectral interferences are known to play a minor role in atomic
fluorescence. With laser excitation, the availability of many non-

Table 1. Detection Limited (ng/ml) obtained by Laser Excited
          Fluorescence Spectrometry[a]

| Element | Detection Limits | |
	Flame[b]	ETA[c]
Ag	4	$3 \times 10{-3}$
Al	0.6	–
Ba	8	–
Bi	3	–
Ca	0.01	–
Cd	8	–
Co	19	2×10^{-3}
Cr	1	–
Cs	–	2×10^{-2} (d)
Cu	1	2×10^{-3}
Eu	20	10
Fe	0.06	10^{-3}
Ga	0.9	–
In	0.2	–
Ir	–	0.2
Li	0.5	–
Mg	0.002	–
Mn	0.4	6×10^{-3}
Mo	12	–
Na	0.1	0.02 (2)[e]
Ni	0.5	–
Pb	0.2	2.5×10^{-5}
Pt	–	4
Sn	3	–
Sr	0.1	–
Ti	2	–
Tl	4	–
V	30	–

a) Values are usually given in aqueous solutions and are calculated
   according to a limiting signal-to-rms noise ratio of 3.
b) Flames used are separated nitrous oxide-acetylene and separated
   air-acetylene burning on conventional slot or capillary burners
   at atmospheric pressure and with pneumatic and ultrasonic nebulization[32].
c) Electrothermal atomizers (graphite cuvettes or carbon rod atomizers)[33].
d) [See Reference 34].
e) Average number of atoms in the probe volume. The atomizer is a
   closed sodium cell (oven)[35].

Table 2. Absolute Detection Limits (fg) Obtained by Laser Excited Fluorescence and Electrothermal Atomization

Element	Detection Limit	
	a)	b)
Ag	100.	–
Bi	–	800.
Co	60.	200.
Cu	150.	800.
Eu	3×10^5	–
Fe	100.	1000.
In	–	100.
Ir	6×10^3	–
Mn	200.	–
Na	600.	–
Ni	–	1000.
Pb	1.5	5.
Pd	–	1000.
Pt	1.2×10^5	–
Tl	–	2.4

a) [See Reference 36].
b) [See Reference 37].

resonance transitions (see Figure 3) should still decrease the possibility of spectral interferences as it was clearly demonstrated in the case of gallium and manganese[9]. However, native flame species (OH, CH, CN, C_2) or molecular species formed in the combustion process can give rise to a fluorescence background[14]. Therefore, it is always advisable to scan the fluorescence spectrum in the vicinity of the analytical line.

Another point of concern for improving the analytical sensitivity is the spot size of the laser in the atomizer. The optimum spot size is of course dictated by the requirement that the collimator should be filled with light. However, one should be aware that much will be gained by enlarging the laser size into the flame, provided that the irradiance is still sufficient to saturate the transition.

Finally, the peak power of the laser should not be overrated in comparison with its spectral quality. Therefore, the use of etalons within the oscillator cavity will result in improved spectral irradiance over the absorption profile with the welcome advantage of better sensitivity and decreased scattering.

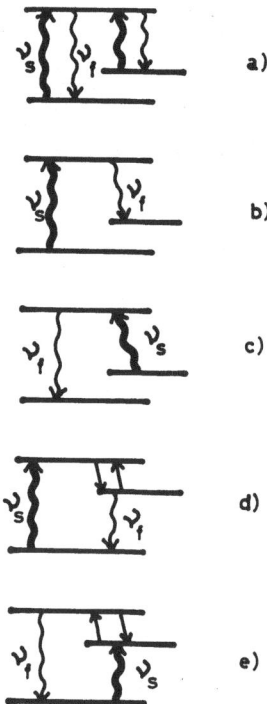

Fig. 3. Several types of fluorescence transitions observed with
laser excitation[5]. ν_s = excitation frequency; ν_f = fluorescence frequency. a) resonance fluorescence; b) Stokes
direct line fluorescence; c) Antistokes direct line fluorescence; d) Stepwise fluorescence; e) thermally assisted
fluorescence.

3. DIAGNOSTIC APPLICATIONS

While the widespread acceptance of the use of lasers in analytical atomic fluorescence spectrometry in flames and plasmas can still
be questioned because of the high cost of the overall apparatus and
the lack of attaining (yet) the ultimate detection limits predicted
by theoretical considerations, it is out of doubt that the technique
of laser induced fluorescence (LIF) has gained a high popularity in
combustion diagnostics with regard to the evaluation of the temperature and total number density of several species produced during the
combustion process. This is well documented in a recent book edited
by Crosley[15].

Like the Raman technique, LIF has the advantage of allowing
local sampling of the parameter sought. This has the advantage over

the conventional absorption and emission methods that no special
techniques such as the Abel inversion are required to convert line of
sight information to radially resolved data. It is worth stressing,
however, that spatial resolution can be obtained even with the absorption and the optogalvanic techniques in flames by crossing two
laser beams, one acting as a weak probe and the other one as the
perturbing beam[16]. One should also be aware that any inhomogeneity
present in the probe volume would be averaged by the detecting system, thereby reducing the spatial resolution. Such inhomogeneities
could be present both in the source (e.g. in the spatial profile of
the laser beam) and in the system investigated (e.g. pronounced
gradients in temperature and in the atomic or molecular densities in
flames and plasmas).

The present discussion is applicable to combustion system at
atmospheric pressure such as all analytically useful hydrogen- and
acetylene-fuelled flames and to various types of plasmas such as the
inductively coupled argon plasma or the microwave plasma. Among the
parameters of interest to analytical chemists as well as to combustion engineers, we restrict our discussion to three of them, namely
the temperature, the number density and the quantum efficiency. The
reader is referred to some selected pertinent literature for the
derivation of the relationships presented and for the details of the
experimental set-up used.

The atomic system considered has 3 energy levels (see Figure 1b)
indicated as 1, 2 and 3 in order of increasing energy (1 is the
ground state).

Several assumptions are made in treating the interaction between
the laser and the atoms (molecules). First, the vapor is assumed to
be optically thin so that no self-absorption is present which would
modify the equations given. Secondly, the laser irradiance is assumed to be homogenous over the entire volume probed by the detection
system and the laser spectral bandwidth is considered to be much
larger than the absorption profile. Finally, coherence effects are
disregarded as well as laser enhanced chemistry and/or ionization,
both of which have been demonstrated to occur in flames at atmospheric pressure[23,24].

3.1. Evaluation of Temperature

The underlying principle of most of the fluorescence methods is
based upon the fact that level 2 is metastable and therefore not
radiatively connected with the ground state, i.e. $A_{21} = 0$ in Figure
1b. Its thermal population is sensed by taking the ratio of two
fluorescence transitions (resonance and antistokes) and assuming
thermodynamic equilibrium in the system.

If the interaction between the laser and the atomic system is linear, we derive the following relationship

$$T_f = \frac{5040 \, E_{12}}{\log \dfrac{E_{\lambda_{23}}}{E_{\lambda_{13}}} + 5 \, \log \dfrac{\lambda_{23}}{\lambda_{13}} + \log \dfrac{B^*_{F_{31}}}{B_{F_{32}}}} \qquad (6)$$

where E_{12} in the energy (eV) of the metastable level, $B^*_{F_{31}}$ is the antistokes fluorescence radiance excited at λ_{23}, and $B_{F_{32}^{31}}$ is the direct line fluorescence radiance excited at λ_{13}. Here, the electro-optical detection system needs to be calibrated and E_λ's measured; however, the knowledge of the transition probabilities is not necessary.

When optical saturation of the transitions is achieved and steady state values are measure, we obtain

$$\frac{B_{F_{32}}}{B^*_{F_{31}}} = \left(\frac{A_{32} \lambda_{13}}{A_{31} \lambda_{23}} \right) \left\{ \frac{1 + \dfrac{g_2}{g_3} + \dfrac{g_1}{g_3} e^{E/kT_f} + \dfrac{g_1}{g_2} e^{E/kT_f} \left(\dfrac{A_{31} + k_{31}}{k_{21}} \right)}{1 + \dfrac{g_1}{g_3} + \dfrac{g_2}{g_3} e^{-E/kT_f} + \left(\dfrac{A_{32} + k_{32}}{k_{21}} \right)} \right\} \qquad (7)$$

and one can see that the knowledge of k_{31}, k_{32} and k_{21} is needed to derive the temperature. Since the values of these constants are difficult to measure in flames at atmospheric pressure, this method is not attractive.

If the fluorescence waveform is temporally resolved and the measurements are taken at the peak, we have

$$\frac{B_{F_{31}}}{B^*_{F_{31}}} = \left(\frac{g_2 + g_3}{g_1 + g_3} \right) \frac{g_1}{g_2} e^{E/kT_f} \qquad (8)$$

where $B_{F_{31}}$ is the resonance fluorescence radiance, excited at λ_{13}. This method is certainly most attractive but has not yet been demonstrated experimentally.

The temperature can also be evaluated from termally assisted fluorescence measurements, i.e. measurements of the fluorescent emission radiated from levels whose energy is higher than that of the level reached by the laser[18]. These higher levels must be populated by thermal collision processes and should reach their steady state values within the duration of the excitation pulse. If the ratio of two fluorescence transitions starting from levels i and j and terminating into the ground state is measured, we obtain

$$
T_f = \frac{\Delta E_{ji}}{\ln \dfrac{A_{igi}}{A_{jgj}} + \ln \dfrac{\lambda_j}{\lambda_i} + \ln \dfrac{B_{Fjl}}{B_{Fil}}}
\tag{9}
$$

Usually, several fluorescence transitions are measured and the temperature is evaluated from the slope of the line obtained by plotting the normalized fluorescence signals versus E_{ij}. With this technique, only one laser is needed.

The above described laser induced fluorescence techniques have been used to obtain relative as well as absolute temperature profiles in flames with high spatial resolution (< 1 mm^3).

3.2. Evaluation of Number Density[1,3,5,15,25-29]

When the optical transition is saturated, an absolute measurement of the fluorescence radiance (Jsec^{-1} cm^{-2} sr^{-1}) permits, for a two-level system, the direct evaluation of n_T. In this case, referring to the nomenclature of the 3-level system considered before, in absence of level 2, we obtain

$$
B_{F_{31}} = \left(\frac{\ell}{4\pi} \right) A_{31} \, h\nu n_T \left(\frac{g_3}{g_1 + g_3} \right)
\tag{10}
$$

where ℓ is the homogenous depth of the fluorescence volume.

When level 2 is taken into consideration, the expression modifies and becomes complicated by the appearance of the (unknown) rate constants. If level 2 is radiatively coupled with level 1, we obtain

$$
B_{F_{31}} = \left(\frac{\ell}{4\pi} \right) A_{31} h\nu n_T \left\{ \frac{1}{1 + \dfrac{g_1}{g_3} + \dfrac{k_{32}}{A_{21} + k_{21} + k_{23}}} \right\}
\tag{11}
$$

If level 2 is radiatively coupled with level 3, we obtain

$$
B_{F_{31}} = \left(\frac{\ell}{4\pi} \right) A_{31} \, h\nu n_T \left\{ \frac{1}{1 + \dfrac{g_1}{g_3} + \dfrac{g_2}{g_3} e^{-E/kT} + \dfrac{A_{32} + k_{32}}{k_{21}}} \right\}
\tag{12}
$$

It is clear that in these cases the <u>direct</u> evaluation of n_T from a saturated fluorescence measurements is not possible.

The same considerations apply to the so-called "<u>slope method</u>", which consists in the plot of the reciprocal of the fluorescence signal versus the reciprocal of the laser power[3,5,7,25,28]. This plot results in a straight line, whose slope contains the quantum efficiency and the total number density. A peculiar outcome of the expression obtained for a 2-level system is that the intercept of the line allows to scale the ordinate values in absolute units[7]. Again, the expressions become more complicated for 3-level systems.

Several measurements in different flames for both atoms and radical species have been reported in the literature[15,25].

3.3. Evaluation of Quantum Efficiency

A direct evaluation of the quantum efficiency of a transition from saturated fluorescence measurements is again possible only for a 2-level atomic system, by deriving from an experimental saturation curve, the parameter E_ν^S, which is called the "<u>stead state saturation spectral irradiance</u>". Indeed, for a 2-level system, E_ν^S is given by the relationship

$$E_\nu^S = \left(\frac{g_1}{g_1 + g_3} \right) \left(\frac{8\pi h \nu^3}{c^3} \right) \left(\frac{1}{Y} \right) \tag{13}$$

where Y is the quantum efficiency, as defined previously.

The steady state saturation spectral irradiance for a 3-level system is still related to the quantum efficiency of the transitions by again the relations contain unknown rate constants. Y can also be evaluated, for a 2-level system, by the slope method described above.

Of course, the most direct evaluation of the overall quantum efficiency of a transition is given by the measurement of the <u>lifetime</u> of the excited level, provided that the radiative transition probabilities to lower levels are known so to allow the calculation of the radiative lifetime τ_{rad}, of the level. Then, if the lifetime τ, is measured, the quantum efficiency is simply given by the ratio (τ/τ_{rad}).

Such measurements have been made in flames[30] and recently in an inductively coupled argon plasma[31].

4. CONCLUSIONS

For analytical applications, the laser induced fluorescence technique has indeed the potential of achieving extremely low limits of detections as it was shown experimentally for electrothermal atomizers at atmospheric pressure and aqueous solutions. The ultimate impressive sensitivities predicted by theoretical considerations have only been achieved in closed vessels and other non-analytical atomizers. In flames and plasmas the sensitivity is yet far from being spectacular because of their inherent background noise and scattering. However, work is still needed in this area to fully characterize the technique. For electrothermal atomizers, the efforts should probably be only directed towards those elements for which the analytical community still demands more sensitivity of the technique.

In the field of combustion chemistry, the fluorescence method appears to be a remarkably attractive diagnostic tool, for both atomic and molecular species. However, because of the highly reactive medium, one should be aware of the many possible deactivation channels that are available to the laser excited atomic or molecular levels, ionization and chemical reactions being two examples of such channels. The interpretation of the experimental data will then require a more thorough investigation of these collateral effects.

REFERENCES

1. C. Th. J. Alkemade, T. Hollander, W. Snelleman, and P. J. T. Zeegers, "Metal Vapors in Flames", Pergamon Press, (1982).
2. C. Th. J. Alkemade, Appl.Spectroscopy, 35:1 (1981).
3. N. Omenetto, ed., "Analytical Laser Spectroscopy", Wiley, New York, (1979).
4. N. Omenetto and J. D. Winefordner, CRC Crit.Rev.Anal.Chem., 13:59 (1981).
5. N. Omenetto and J. D. Winefordner, Progress in Anal.Atom. Spectroscopy, 2:1 (1979).
6. G. M. Hieftje, J. C. Travis, and F. E. Lytle, eds., "Lasers in Chemical Analysis", The Humana Press, New York, (1981).
7. N. Omenetto, P. Benetti, L. P. Hart, J. D. Winefordner, and C. Th. J. Alkemade, Spectrochim.Acta, 28B:289 (1973).
8. C. A. Van Dijk, N. Omenetto, and J. D. Winefordner, Applied Spectroscopy, 35:389 (1981).
9. S. J. Weeks, H. Haraguchi, and J. D. Winefordner, Anal.Chem., 50:360 (1978).
10. N. Omenetto, G. D. Boutilier, S. J. Weeks, B. W. Smith, and J. D. Winefordner, Anal.Chem., 49:1076 (1977).
11. H. Falk, Progress Anal.Atom.Spectroscopy, 3:181 (1980).
12. M. A. Bolshow, A. V. Zybin, and V. G. Koloshinkov, Sov.Quantum Electronic, 7:1808 (1980).

13. R. P. Frueholz and J. A. Gelbwachs, Appl.Optics, 19:2735 (1980).
14. K. Fujiwara, N. Omenetto, J. D. Bradshaw, J. N. Bower, S. Nikdel
    and J. D. Winefordner, Spectrochim.Acta, 34B:317 (1979).
15. D. R. Crosley, "Laser Probes for Combustion Chemistry", ACS
    Symposium Series 134, Washington DC, (1980).
16. J. E. M. Goldsmith, Appl.Physics, B28, 2/3:304 (1982).
17. J. D. Bradshaw, N. Omenetto, J. N. Bower, G. Zizak and J. D.
    Winefordner, Appl.Optics, 19:2709 (1980).
18. G. Zizak, J. D. Bradshaw, and J. D. Winefordner, Appl.
    Spectroscopy, 35:59 (1981).
19. R. J. Cattolica, Appl.Optics, 20:1156 (1981).
20. R. G. Joklik and J. W. Daily, Appl.Optics, 21:4158 (1982).
21. R. P. Lucht, N. M. Laurendau, and D. W. Sweeney, Appl.Optics,
    21:3729 (1982).
22. R. M. Measures, J.Appl.Physics, 39:5232 (1968).
23. C. H. Muller III, K. Schofield, and M. Steinberg, J.Chem.Phys.,
    72:6620 (1980).
24. J. C. Travis, G. C. Turk, and R. B. Green, Anal.Chem., 54:1006A
    (1982).
25. A. C. Eckbreth, P. A. Bonczyk, and J. A. Shirley, "Investi-
    gations of Saturated Laser Fluorescence and CARS Spectro-
    scopic Techniques for Combustion Diagnostics", EPA Report
    600/7-78-104 (1978).
26. J. W. Daily, Appl.Optics, 15:955 (1976).
27. J. W. Daily, Appl.Optics, 16:568 (1977).
28. A. P. Baronawski and J. R. McDonald, J.Chem.Phys., 66:3300
    (1977).
29. G. D. Boutilier, M. B. Blackburn, J. M. Mermet, S. J. Weeks,
    H. Haraguchi, J. D. Winefordner, and N. Omenetto, Appl.
    Optics, 17:2291 (1978).
30. P. Hannaford, Procedings of the XXI CSI, Heyden & Son,
    Cambridge, England, p.250 (1979).
31. H. Uchida, M. A. Kosinski, N. Omenetto, and J. D. Winefordner,
    Spectrochim.Acta B, in press.
32. S. J. Weeks and J. D. Winefordner, (Chapter 8 in Reference 6),
    and J. J. Horvath, J. D. Bradshaw, J. N. Bower, N. S. Epstein
    and J. D. Winefordner, Anal.Chem., 53:6 (1981).
33. M. A. Bolshov, A. V. Zybin, and I. Smirenkina, Spectrochim.Acta,
    36B:1143 (1981)
34. J. P. Hoimer and P. J. Hargis, Jr., Appl.Phys.Lett., 30:344
    (1977).
35. W. M. Fairbank, Jr., T. W. Haensch, and A. L. Schwalow, J.Opt.
    Soc.Am., 65:199 (1975).
36. M. A. Bolshov, A. V. Zybin, and I. I. Smirenkina, Spectrochim.
    Acta, 36B:1143 (1981).
37. J. Tilch, H. J. Pätzold, H. Falk, and K. P. Schmidt, paper given
    at "Analytiktreffen 82", Neubrandenburg, DDR, November 82.

LASER DIAGNOSTICS IN FLAMES BY FLUORESCENCE TECHNIQUES

G. Zizak

CNPM, CNR-Politecnico
Peschiera Borromeo
Milano, Italy

INTRODUCTION

Progress in combustion studies requires the development of laser based diagnostic techniques and their application to different experimental situations. The following examples can give an idea of the problems involved in the understanding and controlling the combustion processes. Pollutants in the exhaust gases have to be measured at trace level and the results of highly sophisticated computer codes, for chemical kinetics and turbulent combustion studies, must be validated by comparison with locally and temporally resolved temperature and species concentration measurements.

For these purposes laser excited fluorescence (LEF) is a well suited technique as it is spatially and temporally precise, very sensitive and capable of detecting trace species. In past years several experimental methods have been proposed and tested in laboratory flames to perform temperature and concentration measurements. A great deal of work is now in progress to extend the applicability of the LEF technique to large scale and industrial combustion systems. In this paper two techniques for temperature and concentration measurements will be described and some experimental results will be also reported.

TEMPERATURE MEASUREMENTS

Among the five different LEF techniques that have been proposed[1], the thermally assisted fluorescence technique[2] seems to have the most to offer for spatial and "instantaneous" temperature measurements with commercially available equipment. Basically the technique can be described by considering Figure 1 where a multilevel (atomic or molecular) system, with collisional and radiative (when present) rates R, is depicted. The two laser coupled levels are the ground state g, and an upper level e. The laser enhanced population of level e, n_e, is partially transferred to other levels j by collisions with the surrounding gases. The steady state rate equations governing the population densities of an individual level i (in the manifold of levels j) and of the manifold of levels j are:

$$n_e R_{ei} + n_g R_{gi} + \Sigma_j n_j R_{ji} - n_i (R_{ie} + R_{ig} + \Sigma_j R_{ij}) = 0 \tag{1}$$

$$n_e \Sigma_j R_{ej} + n_g \Sigma_j R_{gj} - \Sigma_j n_j (R_{je} + R_{jg}) = 0 \tag{2}$$

In the laser excited fluorescence experiments we can safely neglect the upward collisional transfer rates ($R_{ge} = R_{gj} = R_{gi} = 0$). Calling:

$$n_c = \Sigma_j n_j, \qquad\qquad R_{ec} = \Sigma_j R_{ej}, \qquad\qquad R_{ic} = \Sigma_j R_{ij}$$

$$R_{ci} = (\Sigma_j n_j R_{ji})/\Sigma_j n_j, \; R_{ce} = (\Sigma_j n_j R_{je})/\Sigma_j n_j, \; R_{cg} = (\Sigma_j n_j R_{jg})/\Sigma_j n_j$$

and considering the case of a collision dominated system, we have[3,4]

$$n_c/n_e = \Sigma_j n_j/n_e = (n_c/n_e)_B \cdot R_{ce}/(R_{ce} + R_{cg}) \tag{3}$$

$$n_i/n_c = (n_i/n_c)_B \cdot \frac{(R_{ie} + R_{ic})/R_{ig} + (R_{ie}/R_{ig})/(R_{ce}/R_{cg})}{(R_{ie} + R_{ic})/R_{ig} + 1} \tag{4}$$

where the subscript B means the Boltzmann ratio, given only by the temperature of the system. The ratio $(R_{ie} + R_{ic})/R_{ig}$ is the ratio of the excited state energy transfer and the quenching to the ground state; this ratio is frequently called the branching ratio[5]. From Equation (3) we see that for very high values of the branching ratio ($R_{ce} \gg R_{cg}$) the collisional excited levels are in Boltzmann equilibrium with the laser excited level, but as a general rule, a deviation is present and level e is overpopulated.

Equation (4) describes the population redistribution within the manifold. Levels for which the ratio $r = (R_{ie}/R_{ig})/(R_{ce}/R_{cg}) = 1$ are in equilibrium with the manifold. The other levels are overpopulated or underpopulated according to Equation (4). However the deviation from the Boltzmann distribution is a weak function of the ratio r because the collisional mixing in the upper levels is, in general, greater than the quenching. Knowledge of the collisional rates

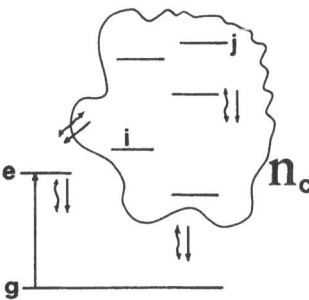

Fig. 1. Scheme of a multi-level system under external irradiance
         with radiative and collisional decay rates.

allows to calculate the temperature of the gases by fitting the
population distribution measured by fluorescence experiments with
Equations (3) and (4).

Moreover for levels sufficiently higher than the laser excited
level ($R_{ie} \ll R_{ig}$) Equation (4) can be reduced to

$$n_i/n_c = (n_i/n_c)_B \cdot \frac{1}{(1 + R_{ig}/(R_{ic}+R_{ie}))} \qquad (5)$$

For these levels, with good approximation, the deviation from the
equilibrium distribution can be considered constant. Hence a plot of
the population densities vs. the level energy shows a straight line
with slope given by the temperature of the system. In Figure 2 a
qualitative picture of the typical population redistribution in LEF
experiments is depicted.

Several flames have been tested with the thermally assisted
technique using a nitrogen laser pumped dye laser, a conventional low
resolution receiving optics, a fast photomultiplier and a boxcar
averager[2,3,4]. As an example the results obtained by seeding an
acetylene-oxygen-argon flame with Tl atoms are reported. A simplified energy diagram of Tl is given in Figure 3 with the fluorescence
transitions observed being denoted.

Figure 4 shows the Boltzmann plot of the level populations.
Lines coming from the same level, in absence of post-filter or polarization effects, should fall on the same point. Verification of that
confirms the reliability of the measurements, the accuracy of the
A-values employed and the accuracy of the calibration of the optical
system. The $7S_{\frac{1}{2}}$ level, directly excited by the laser tuned at 377.6
nm, is overpopulated, while the 6D, 8S and 7D levels are in Boltzmann
equilibrium corresponding to a temperature of 2498 K with an estimated
error of $\pm$ 50°K. Pumping the transition $6P\,^{1}/_{2} \rightarrow 6D\,^{3}/_{2}$, we observe

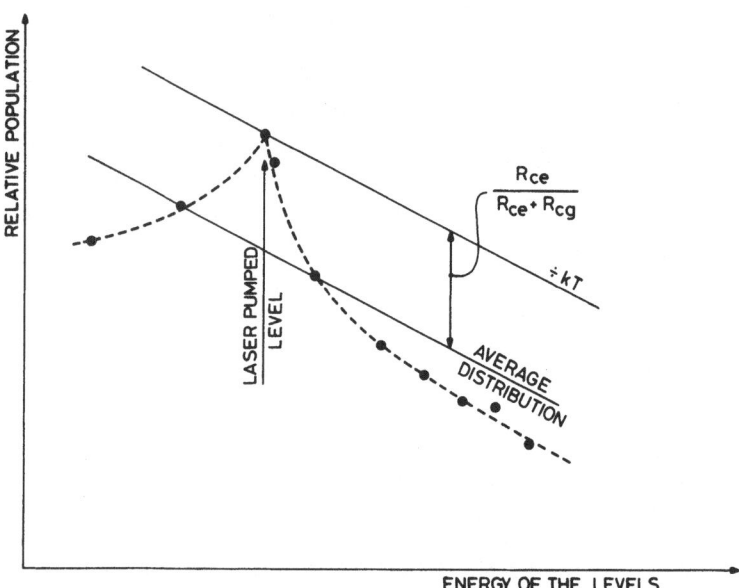

Fig. 2.   Qualitative picture of the relative population redistri-
          bution in laser excited fluorescence.

that, again the two levels above are in Boltzmann equilibrium with
respect to each other, that the laser excited level is overpopulated
but the $7S_{\frac{1}{2}}$ level is populated by downward collisions from the $6D_{\frac{3}{2}}$
level. The line reversal temperature, obtained by seeding 1000 ppm
of Na in the flame, resulted to be (2470 ± 15°K), in good agreement
with the value calculated from the fluorescence measurements.

    The same concept of the thermally assisted technique has been
applied to the OH radical present in a stoichiometric methane-air[6].
A flash lamp dye laser was used to excite the Q_13 transition at 308.2
nm of the $\nu(0,0)$ vibrational band and the fluorescence spectrum was
detected through an 1.0 m monochromator. Figure 5 shows the population redistribution within the rotational manifold. Levels that
are one kT higher than the laser excited keep a Boltzmann equilibrium
among them, corresponding to a temperature of 2020°K in very good
agreement with recent measurements in methane-air flames[7]. The
same distribution has been found also exciting the Q_13 transition by
a nitrogen pumped dye laser with 5 nsec pulses. This result indicates that steady state conditions can be reached for the rotational
relaxation of OH molecules in atmospheric pressure flame during very
short laser pulses.

    The main advantages of the thermally assisted technique can be
summarized as follows:

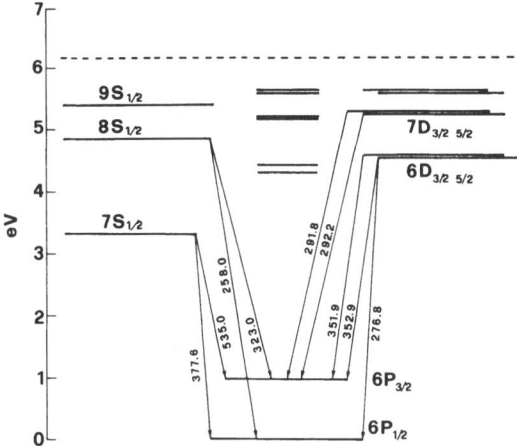

Fig. 3. Thallium partial energy diagram with the levels and the transitions of interest [redrawn from Reference 3].

1) Simplicity: only one laser wavelength is needed. For "instantaneous" measurements two photomultipliers or a multi-channel optical detector must be employed.
2) The temperature measurements are independent of the source spectral irradiance. Hence corrections for the spatial, temporal and spectral profiles of the pulses are not needed.
3) Corrections for pre-filter effects are not needed due to point 2.
4) Corrections for post-filter effects are negligible, since the thermometric species is a trace constituent in comparison with the major species of the combustion gases (thin approximation) and because transitions from the collisional populated levels to lower excited levels can be chosen for detection.
5) The influence of Mie and Rayleigh resonance scatter radiation can be made negligible since lines of different wavelength can be chosen.

The disadvantages present are:

1) Knowledge of the A values of the transitions must be available.
2) The electro-optical set-up must be calibrated.
3) The system must reach a steady state condition during the laser pulse, to allow a complete redistribution of the population.

CONCENTRATION MEASUREMENTS

The measurement of ppm level species concentration in combustion environments can be performed by saturated laser-excited fluorescence technique. From the rate equation approach and the steady state approximation it may be shown that the number density of the laser

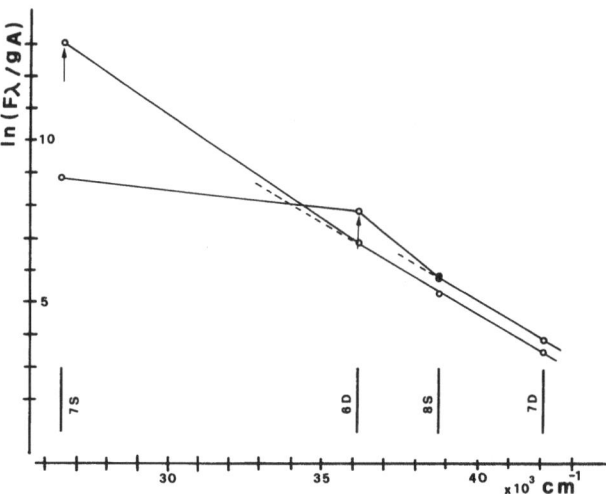

Fig. 4. Boltzmann plot of the Tl level populations. The laser
excited levels are overpopulated, while the levels above
keep a Boltzmann equilibrium corresponding to the flame
temperature [redrawn from Reference 3].

excited level n_e is related to the total number density n_{tot} by[8,9].

$$n_e = n_{tot} \cdot (1+g_g/g_e+R_{ec}/(R_{ce}+R_{cg}))^{-1} \cdot (1+\rho_{sat}/\rho)^{-1} \qquad (6)$$

where g_g and g_e are the statistical weights of the lower and upper
levels connected by the laser and ρ_{sat} is the saturation spectral
density:

$$(7)$$

$$\rho_{sat} = (R_{ec}+R_{eg}-R_{ce}R_{ec}/(R_{ce}+R_{cg})) \cdot B_{ge}^{-1} \cdot (1+g_g/g_e+R_{ec}/(R_{ce}+R_{cg}))^{-1}$$

The saturation spectral density depends on the collisional constants
and the Einstein transition probability for absorption B_{ge}. It is
well known that for very high laser irradiances, when $\rho_{sat} \ll \rho$, the
transition saturates and the fluorescence signal reaches a maximum.
In a two-level system the total number density can be easily measured
from the saturated fluorescence signal since $R_{ec} = R_{ce} = 0$, but for a
three-level or a multi-level system knowledge of the collisional
rates is needed. However in the case of sodium-like atoms, from a
set of measurements, the $R_{ec}/(R_{ce} + R_{cg})$ ratio can be calculated and
the concentration can be obtained[10].

Sodium atoms are considered as three level systems, composed by
the $3P_{\frac{1}{2}}$ ground state 1, and by two upper states 2 and 3 corresponding
to the $3S_{\frac{1}{2}}$ and $3S_{\frac{3}{2}}$ levels. Exciting the $3P_{\frac{1}{2}} \rightarrow 3S_{\frac{3}{2}}$ transition,

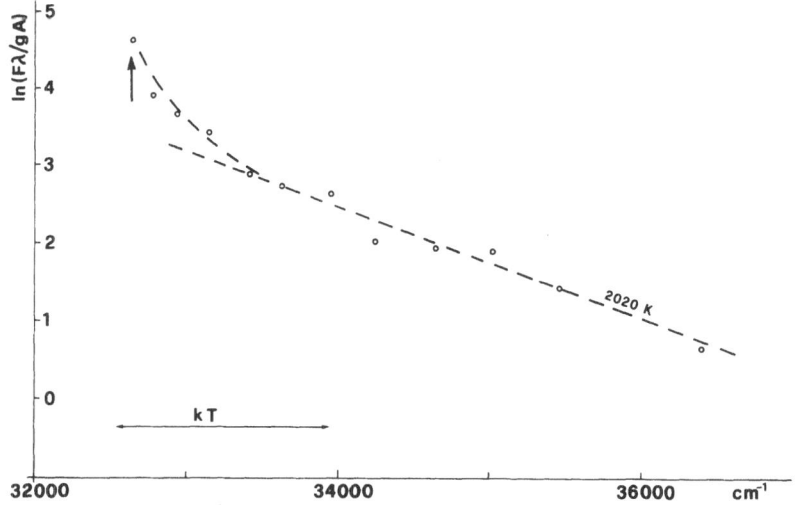

Fig. 5. Boltzmann plot of the OH rotational population. The arrow
         indicates the directly laser-excited level [redrawn from
         Reference 6].

measuring both the D_1 and D_2 fluorescence lines and remembering
Equation (3), we can write:

$$(n_3/n_2)_{D_2} = (n_3/n_2)_B + (k_{21} + A_{21})/k_{32} \tag{8}$$

For pumping the $3S_{\frac{1}{2}} \rightarrow 3P_{\frac{1}{2}}$ transition we have:

$$(n_3/n_2)_{D_1} = (n_3/n_2)_B \cdot k_{32}/(k_{32} + k_{31} + A_{31}) \tag{9}$$

where B means the Boltzmann ratio, the A's are the Einstein transition probabilities for spontaneous emission and k_{ij} is the collisional rate for the transition from level i to level j. In the
case of complete saturation from Equation (6) and from the collisional rates ratio given by Equations (8) and (9), the concentration of sodium atoms can be calculated:

$$n_{tot} = n_{3max} (1 + g_1/g_3 + (n_2/n_3)_{D_2}) \tag{10}$$

$$n_{tot} = n_{2max} (1 + g_1/g_2 + (n_3/n_2)_{D_1}) \tag{11}$$

where $n_{e\ max}$ is the saturated population of the laser excited level.
Moreover is, experimentally, complete saturation cannot be reached, a
slope method can be applied[11,12]. Equation (6) shows that, in the
region of near saturation, a plot of the reciprocal of the fluorescence signal, measured in absolute units, vs. the reciprocal of the
laser intensity exhibits straight line behavior and intercept corres-

ponding to the absolute maximum concentration of the laser excited level $n_{e\ max}$. Hence, from the inverted saturation curve, the total number density can be calculated using Equations (10) and (11).

Sodium atoms were introduced into an acetylene-air flame by pneumatic nebulization of a 20 ppm water solution. Flame temperature (2320°K) by line-reversal method and average sodium concentration (30.0 10^9 atoms/cm^3) by absorption technique measurements have been performed using a calibrated tungsten strip lamp. A cw broad band dye laser, weakly focused at the center of the flame, has been used to excite the sodium fluorescence. The detection system was composed by a 20 cm monochromator, a high gain photomultiplier with associated photon-counting electronics. Four saturation curves, corresponding to the cases of D_1 and D_2 line excitation and detection were measured with the aid of calibrated neutral density filters.

In Figure 6 the saturation curves for D_2 excitation and D_1 and D_2 detection are reported. It is apparent that complete saturation has not been reached and the slope method has to be applied to calculate the sodium concentration. The $(k_{21} + A_{21})/k_{32}$ and the $(k_{31}+ A_{31})/k_{32}$ rates, obtained from the saturation curves and compared with other previously published values, are reported in Table 1[13,14]. Because A_{21} and A_{31} are equal, the measurements indicate a difference in the quenching rates of level $3S_{\frac{1}{2}}$ and $3S_{\frac{3}{2}}$ but not as large as was claimed[13]. Figure 7 shows the inverted saturation curve and its

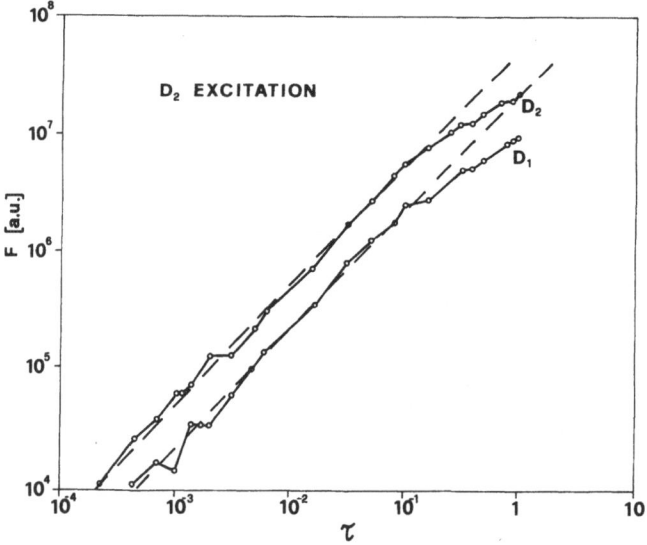

Fig. 6. Saturation curve of the D_1 and D_2 lines of sodium under cw broad band laser excitation. F is the fluorescence signal and τ is the filter transmissivity.

Table 1. Collisional Rates Ratio for the Sodium Doublet in the
          Acetylene-Air Flame

	$(k_{21} + A_{21})/k_{32}$	$(k_{31} + A_{31})/k_{32}$
This work	0.341 ± 0.193	0.475 ± 0.130
Allen[13]	0.120 ± 0.030	0.478 ± 0.033
Omenetto[14]	0.299 ± 0.151	0.467 ± 0.065

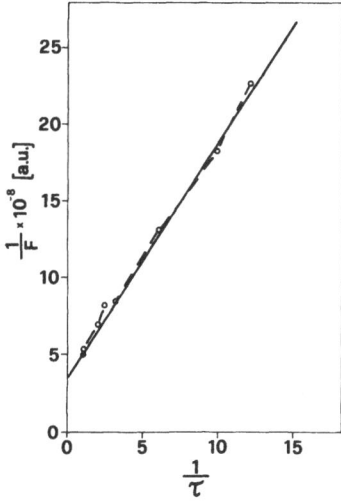

Fig. 7. The inverted saturation curve from the data of Figure 6.

linear behavior. From the intercept, using the appropriate g's, the
measured n_2/n_3 value and the calibration factor, the sodium concentration was estimated to be $2.7 \cdot 10^9$ atoms/cm^3, that is one order of
magnitude less than the absorption measured value.

     Several considerations are possible. The saturation plateau was
not reached because of the gaussian shape of the beam[9,15]. Taking
into account this fact would lower the number density measured by
fluorescence. Moreover fluorescence is a spatial resolved technique
while absorption measurements are averaged through the flame. A very
careful calibration is also needed for precise measurements. However
it is evident that depletion of sodium atoms occurs in cw laser
excited fluorescence experiments. It has been found that laser
enhanced ionization (LEI) and laser enhanced chemistry (LEC) effects
drive atoms out of the laser excited levels to some metastable states
(lifetime > 1 µsec) that cannot be detected by optical methods. As a
result a lower ground state population is measured[16]. Instead,
using short laser pulses (nsec range) with good beam quality, the

slow rate depletion effects (typically LEC) have no time to occur and
reliable concentration measurements can be performed.

Acknowledgements

The author wishes to thank Prof. J.D. Winefordner and his research group from University of Florida, Mr. S. Benecchi and Mr. F.
Cignoli from CNPM, Milano for helping him during the experiments
reported in this work.

REFERENCES

1. J. D. Bradshaw, N. Omenetto, G. Zizak, J. N. Bower, and J. D.
    Winefordner, Five laser excited fluorescence methods for
    measuring spatial flame temperatures. 1: Theoretical basis,
    Appl.Opt., 19:2709 (1980).
2. G. Zizak, J. D. Bradshaw, and J. D. Winefordner, Thermally
    assisted fluorescence: A new technique for local flame tem-
    perature measurements, Appl.Spectrosc., 35:59 (1981).
3. G. Zizak, J. J. Horvath, C. Van Dijk, and J. D. Winefordner,
    Collisional redistribution of radiatively-excited levels of
    Tl and Ga atoms in an O_2-acetylene-Ar flame, J.Quant.
    Spectrosc.Radiat.Transfer, 25:525 (1981).
4. G. Zizak and J. D. Winefordner, Application of the thermally
    assisted atomic fluorescence technique to the temperature
    measurement in a gasoline-air flame, Comb.Flame, 44:35
    (1982).
5. C. Chan and J. W. Daily, Laser excitation dynamics of OH in
    flames, Appl.Opt., 19:1357 (1980).
6. G. Zizak, J. J. Horvath, and J. D. Winefordner, Flame tempera-
    ture measurement by redistribution of rotational population
    in laser excited fluorescence: An application to the OH
    radical in a methane-air flame, Appl.Spectrosc., 35:488
    (1981).
7. R. J. Cattolica, Laser absorption measurements of OH in a
    methane-air flat flame, SAND 79-8717.
8. N. Omenetto, P. Benetti, L. P. Hart, J. D. Winefordner, and
    C. Th. J. Alkemade, Non-linear optical behavior in atomic
    fluorescence flame spectrometry, Spectrochim.Acta, 28B:289
    (1973).
9. J. W. Daily, Saturation of fluorescence in flames with a
    gaussian laser beam, Appl.Opt., 17:225 (1978).
10. M. B. Blackburn, J. M. Mermet, G. D. Boutilier, and J. D.
    Winefordner, Saturation in laser excited atomic fluorescence
    spectrometry: experimental verification, Appl.Opt., 18:1804
    (1979).
11. N. Omenetto and J. D. Winefordner, Atomic fluorescence spectro-
    metry: Basic principles and applications, Prog.Anal.Atom.
    Spectrosc., 2:1 (1979).

12. A. P. Baronavski and J. R. McDonald, Application of saturation spectroscopy to the measurements of C_2, $^3\Pi_u$ concentrations in oxyacetylene flames, Appl.Opt., 16:1897 (1977).

13. J. E. Allen, W. R. Anderson, D. R. Crosley, and T. D. Fansler, Energy transfer and quenching rates of laser-pumped electronically excited alkalis in flames, 17th Symp. (Internat.) on Combustion, pp.797, The Combustion Institute (1979).

14. N. Omenetto, M. S. Epstein, J. D. Bradshaw, S. Bayer, J. J. Horvath, and J. D. Winefordner, Fluorescence ratio of the two D sodium lines in flames for D_1 and D_2 excitation, J.Quant. Spectrosc.Radiat.Transfer, 22:287 (1979).

15. L. Pasternack, A. P. Baronavski, and J. R. McDonald, Application of saturation spectroscopy for measurement of atomic Na and MgO in acetylene flames, J.Chem.Phys., 69:4830 (1978).

16. M. Iino, H. Yano, Y. Takubo, and M. Shimazu, Saturation characteristics of laser induced Na fluorescence in a propane-air flame; the role of chemical reaction, J.Appl.Phys., 52:6025 (1981).

ANALYTICAL SPECTROSCOPY USING LASER ATOMIZERS

K. Laqua

Institut für Spektrochemie und angewandte Spektroskopie
Dortmund
Federal Republic of Germany

INTRODUCTION

In this article, laser as a source of thermal energy supplementing customary spectrochemical excitation sources will be treated. It is therefore necessary to first explain the properties of typical excitation sources. In optical spectrochemical analysis, chemical elements contained in a sample are determined with the help of their optical line spectra. This can only be done, if a representative part of a solid or liquid sample is first converted into atomic vapor. Free atoms can then either be determined with the help of their absorption spectra or, if suitably excited, with the help of their emission spectra. It is, of course, also possible to detect the elements with the help of ionized atoms either by atomic emission spectroscopy or by mass spectrometry.

The customary radiation sources can be divided into two groups, e.g. electrical radiation sources and non-electrical radiation sources. The former group comprises electrical arcs, sparks and discharges under reduced pressure or in vacuum. Evaporation of sample material and excitation are both effected by the same source. Electrically conducting samples can be analysed directly, whereas electrically non-conducting samples have to be suitably prepared e.g. by grinding them down and mixing with conducting material. It is well known that the evaporation and excitation in electrical radiation sources can be strongly influenced by the chemical and physical properties of the sample. Great care has therefore to be exercised to avoid systematic analytical errors.

The second group of radiation sources can best be called 'plasmas'. They comprise chemically produced plasmas, e.g. flames

159

and electrically produced plasmas, e.g. microwave induced plasmas,
induction coupled plasmas, etc. Commonly, the sample has to be
converted into an aerosol which then is introduced into the plasma
and excited. Solid materials are usually first dissolved, but direct
production of an aerosol from such substances is also possible.

   For local analysis, several methods have been described in
literature. With electrical radiation sources, spatial resolution
can be improved by using a pointed counter electrode, preferably in
connection with a confinement of the surface area which can be attacked by the discharge. With non-conducting surfaces, a local
analysis can only be performed indirectly by first mechanically
removing the respective part of the sample and subsequently analysing
it by a standard method.

   Soon after the first realization of a Ruby laser in 1960 by
Maiman[1] in 1962, a 'Laser Microprobe' designed for local analysis
was described and introduced into analytical routine by Brech and
Cross[2]. Understandably, due to lack of fast and powerful methods
for local spectrochemical analysis, the first methods of application
of laser centered around this special analytical technique. Since
then, the scope of application has widened considerably and now
comprises surface, micro- and macro-analysis as well. The particular
properties of laser as excitation source in optical and mass spectroscopy rest upon the fact that powerful radiation is emitted with a
very small angle of divergence. It can therefore, be concentrated
into a very small spot at the focus of the lens. The irradiance of
material in this spot may be sufficiently high for vaporizing material for spectral analysis. This applies to all kinds of materials
irrespective of their electrical properties. The vapor may have such
a high temperature, depending on the mode of operation of the laser
system, that the characteristic radiation of the pertaining atoms is
excited or the atoms are ionized. It is also possible to further
excite the laser-produced vapor plume by several means. Both methods
have their relative merits. The principle of operation of a laser
source for optical spectroscopy can best be understood with the help
of Figure 1, which is a semi-schematic representation of a commercial
instrument. It consists basically of the laser including the Q-
switch, a microscope for aligning the sample and focusing the laser
radiation on the surface and an electrode system for auxiliary cross
spark excitation. The instrument as shown, is primarily thought for
optical emission spectroscopy in local analysis, but with some alter-
ations, its applicability could be extended. A general survey of the
analytical facilities obtainable with the help of laser atomizers is
given in Figure 2. A large number of research papers dealing with
principles and applications of laser atomizers have been published.
They are reviewed in articles by Scott and Strasheim[3] (150 references), Laqua[4] (113 references), Petukh and Yankovskii[5] (219
references). The book by Moenke and Moenke-Blankenburg[6] should
also be consulted. An assessment of quantitative analytical possi-

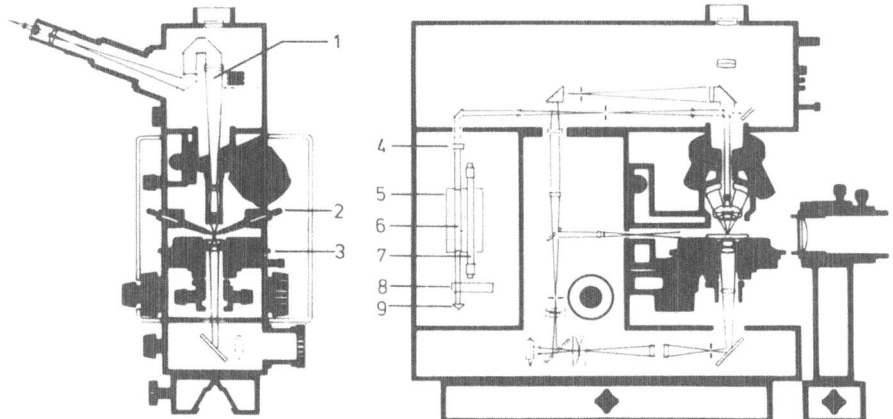

Fig. 1. Laser Micro Analysator LMA 10, Left, front view, side view,
1) observation microscope, 2) electrodes for cross discharge, 3) microscope stage with sample holder, 4) semi-
transparent resonator mirror, 5) laser cavity, 6) laser rod,
7) flash-lamp for optical pumping, 8) liquid dye Q-switch,
9) resonator prism. [VEB Carl Zeiss, Jena].

bilities is given by Van Deijck, Balke and Maessen[7]. Other more
recent publications are the following[8-60]. Laser-induced formation
of ions for mass spectrometry has first been described by Honig and
Woolston[61] in 1963. Since then, numerous reports on this subject
have been published. The most recent, thorough review is that by
Concemius and Capellen[62] (462 references). In addition, the new
work by Jansen and Witmer[63,64] should be mentioned demonstrating
the present state of the art. The history of high resolution laser
microprobe mass spectrometry has been dealt with by Denoyer, Van
Grieken, Adams and Natusch[65] who then describe in detail, a commercial instrument for this purpose, based on the work by Hillenkamp,
Kauffmann, Nitsche and Unsöld[66].

LASERS AS USED FOR ATOMIZATION

Some lasers suitable for atomization of solid samples and their
operation ranges are listed in Table 1. Optically pumped as well as
electrically excited configurations are used. Liquid lasers, although listed, have hardly been employed. By far the widest application have found the solid-state lasers, but it can be expected that
excimer lasers, especially rare gas halogen lasers may have a future.
For particulars concerning the latter, the reader is advised to
consult for instance, the report by Hutchinson[67].

Table 1. Some Properties of Lasers suitable for Atomization of Solid Samples

Type of laser	Wavelength λ μm	Excitation	Output energy of free-running laser, J	Range of power per pulse, MW	Angle of divergence mrad	Repetition rate pps
Solid state						
Ruby	0.6943*	optically pumped	10	0.1–10	3–5	low
Nd-glass	1.06 *		10	0.01–10	5	low
Nd: YAG	1.06 *		10	0.01– 0.1	3–5	10^3
Fluid state						
Nd: POCl$_3$	1.06	optically pumped	5	~10	~10	10
Nd: SeOCl$_2$	1.06					
Dyes	0.22–0.74		1	~ 0.1	2	100
Gaseous state						
CO$_2$-TEA	10.6	electrically excited	100	~ 1	5	100
N$_2$	0.337		0.01	~ 0.1	~10	100
Excimer Oscillator	0.193–0.351	electrically excited	~0.2	~20	2x4	100
Excimer Oscillator + Amplifier	0.193–0.351	electrically excited	~0.3	~15	0.2	40

*also used at λ/2, λ/3 and λ/4 obtained by frequency multiplication.

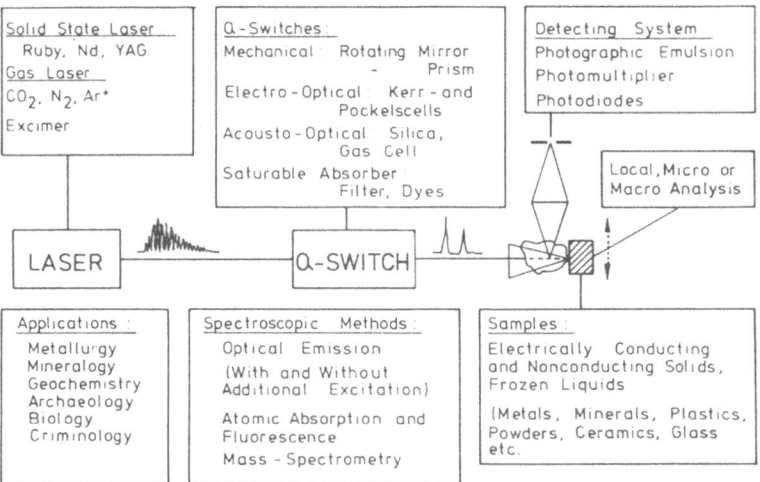

Fig. 2. Analytical possibilities in combination with laser atomizer.

With solid-state lasers, a wide range of output energy and power
will be available depending on the size of the laser rod and on the
pumping system. Rod lengths from 15 to 150 mm are typical with
diameters ranging from 5 to 15 mm. The output energy may be between
a few tenths of a joule and about 5 joule, but applications with
considerably higher energy have been reported. Typical properties of
ruby as active medium are the following: radiation in the visible
region (0.6943 μm), good heat conductivity, comparatively high threshold energy, small angles of divergence with high quality rods.
Glass rods doped with Nd have lower thresholds, radiate in the near
infrared (1.06 μm) which may be advantageous for the analysis of
dielectric materials, and are thermally more sensitive than ruby
rods. Laser with rods fabricated from yttrium aluminium garnet (YAG)
doped with neodymium permit very low lasing thresholds and therefore,
are very suitable for higher repetition rates. Their peak power
capability however, is more restricted than that of the other two.

In the free running operation mode, optically pumped lasers,
once threshold is passed, radiate in a random sequence of single
pulses called spikes which may have powers of a few kW. An emission
of this kind is indicated in Figure 2. For operation of the laser
system in a Q-switched mode, several types of Q-switches may find
application. The simplest type of a Q-switch is a bleachable filter
which may either be solid or liquid. With such switches, one to
several spikes per pumping cycle can be produced possessing powers of
up to several MW. It should be mentioned, however, that this Q-
switch cannot be externally triggered, a property which may be desirable for certain applications. Kerr cells or Pockels cells, very
common with other high power laser operation, are also used in an-

alytical practice. Rotating slotted disks, mirrors or prisms as
Q-switches permit the operation with analytical properties lying
between those of the free running and the fast Q-switched laser.
Several types of acousto-optical Q-switches have been described, one
of which produces a series of equidistant spikes for a considerable
fraction of the pumping time, having powers of several 100 kW. This
mode of operation is useful for some spectrochemical applications.

Very short, high power pulses can be obtained by mode locking,
i.e. by forcing all modes in a laser oscillator to oscillate in
phase, and with similar amplitudes in place of the random emission of
spikes due to multimode oscillation. It has been claimed that mode-
locked operation can be used with advantage with laser atomizers[12].

INTERACTIONS OF LASER WITH SOLIDS

Radiation from a laser source is emitted with a very small angle
of divergence (see Table 1). Its radiant intensity, that is, its
radiant power per solid angle, can be very high, exceeding that of
all other known sources. Focusing the laser radiation onto the
target with the help of a lens or mirror can be very effective in
producing small focal spots with corresponding high irradiance sufficient to vaporize any kind of material. By this interaction a
crater is formed. Its dimensions depend on the area of the focal
spot, on the mode of operation of the laser system, on the total
energy delivered to the sample within the focal spot and on properties of the sample. The diameter of the crater does not coincide
with the diameter of the focal area. In Figure 3 the influence of
some of the factors is demonstrated.

The mass of metal eroded by a free running laser is directly
proportional to the fraction of absorbed energy and inversely proportional to the energy required to melt unit mass of the material[17].
Regarding the crater shape it can be said that the depth of the
crater increases more rapidly with increasing energy than the diameter. At constant energy, the shape of the crater depends strongly
on the mode of operation. An unswitched laser emitting a large
number of spikes of low power and low energy produces a rather deep
crater in relation to its diameter due to the subsequent action of a
large number of spikes following at very short intervals in which the
material does not solidify. Usually, a wall of molten material is
forming around the crater. Little overheating action takes place and
the temperature of the vapor cloud is usually rather low with the
consequence of very little emission of radiation. A considerable
part of the material is not atomized either, but is ejected in the
form of small liquid or recondensed particles. The fraction of vapor
is of the order of 0.1%[17]. As this mode of operation is capable of
producing the smallest craters (there is a threshold below which no
evaporation takes place), it is most widely used in local analysis,

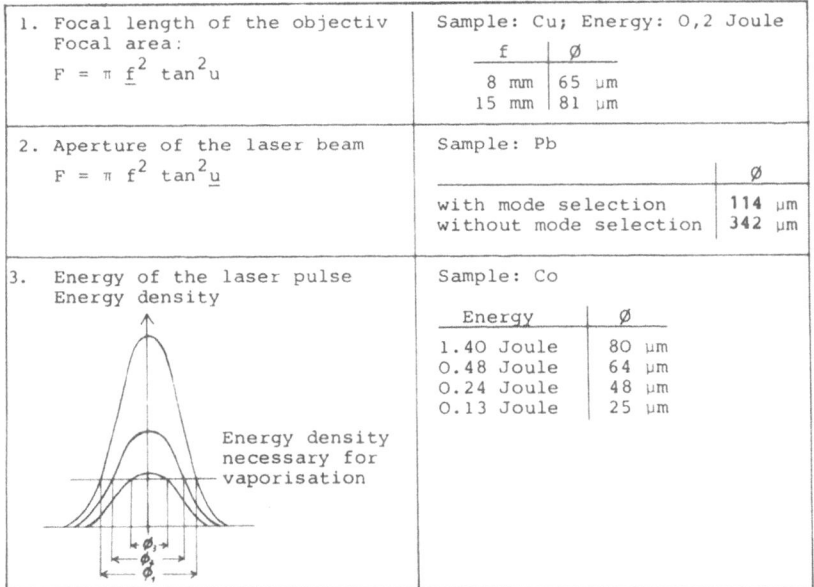

Fig. 3.   Factors affecting diameter Φ of laser-produced crater, in a
          solid sample.

but because of the poor radiation efficiency, additional excitation
of the vapor cloud is compulsory.

   The other extreme is the action of one single giant pulse as
obtainable, for instance, with the help of a Kerr or Pockels cell.
With this mode of operation, very high power and power densities on
the surface can be obtained.   The result of the interaction with the
surface is a very shallow crater with little or no molten residue.
The diameter may be rather large.   Because of the rapid overheating
of sample material, the initial temperature of the vapor cloud expanding rapidly at initial velocities of a km per second or more, is
high.   From the appearance of the crater, it can be concluded that
most of the removed material is atomized with only a very low fraction of molten or solid particles.   The vapor cloud may radiate
strongly.   At the beginning, the radiation consists of a strong
background due to the high pressure and rather broad spectral lines.
At a later stage, the background is reduced and the spectral lines
become narrow; later still, band spectra are also detectable.   The
atomic radiation can directly be used for spectrochemical analysis at
not too low concentrations.

   With the help of bleachable filters it is possible to obtain
several giant pulses at rather large intervals (distributed over most
of the pumping time).   The crater, due to the repeated attack of

laser radiation on the sample surface, is deeper but otherwise of
similar appearance because each giant pulse hits a rather cool surface. The spectra are very similar to those described above.

With Q-switch operation employing acousto-optical switches as
described above, the dimensions of the crater and its shape are
in-between those obtainable with the free running laser and those
obtainable by a sequence of giant spikes. Each of the 50 to 100
spikes interacts with a solid and rather cool surface. On the other
hand, as compared to the free running laser, the power of the spikes
is sufficiently high and consequently, the temperature of the vapor
cloud, to effectively excite the atomic spectra of the respective
elements, but with considerably less background.

This mode of operation should therefore be used for the direct
analysis of solids in cases where high spatial resolution (very small
crater) and high power of detection are not required. Regarding the
erosion of material from solid surfaces, some more generalization can
be made: with samples of low melting points, more material is removed, mostly in the form of liquid droplets, than predicted by
theory. Crater diameters can be less reliably forecast than crater
depths. The reflection factor of the sample surface may also be of
influence, but mainly at comparably low power. Special attention has
to be paid if transparent or little absorbing materials have to be
analysed, especially if they have high boiling temperatures, e.g.
glasses. In these cases a high irradiance (10^9 W cm^{-2}) is necessary.
Plastic material behaves similar to metals of low melting boiling
points. Comparably little atomic vapor is produced and the craters
are rather large. Liquid materials can also be analysed but it is
necessary to first convert them to solids by freezing. The same
technique has been applied with success to biological samples, e.g.
tissues.

On principle, gases could be handled in the same way although no
such investigations have become known. They also could be analysed
directly with the help of 'laser sparks', i.e. the direct dielectric
breakdown in the respective gas caused by high power focused laser
radiation. Analytical application of that technique apparently has
also not been undertaken yet.

By strict control of the laser operating parameters, especially
pumping energy and temperature, the reproducibility of the laser
output can be high, whereas the resulting crater dimensions vary much
more, typically 10% as in comparison to 1%. It has been shown that
the reproducibility of crater dimensions can be significantly improved in cases where a face of a single crystal is attacked.

Vul'fson et al.,[10] have investigated in detail the vaporization of material in a laser-produced plume, especially with the aim
to ensure conformity of composition of the vapor with the composition

of the sample. For a more thorough treatment of the subject of
interaction of laser radiation with solids, the reader is referred to
the book by Ready[69]. Pertinent original papers are discussed for
instance, in[3,4,5,13,17,40,62].

ATOMIC ABSORPTION SPECTROSCOPY USING LASER ATOMIZERS

The laser-produced vapor cloud contains a considerable propor-
tion of neutral ground state atoms. Therefore, atomic absorption
methods can be used to identify the elements producing these atoms.
This has first been employed by Mossotti et al.,[70]. The analytical
signal must be separated from interfering emission signals which are
usually also generated in the vapor cloud and should be corrected for
nonspecific absorption in a standard manner. The former can be
accomplished by choosing a suitable distance from the sample. This
situation is illustrated in Figure 4.

Basically, three different approaches for designing atomic
absorption methods can be distinguished, namely those in which the
atomic absorption is directly observed, those in which the vapor
cloud is in situ subjected to additional heating and those in which
the vapor is transferred to the final atomization device, which may
be a flame, a plasma or a furnace. Figure 5 shows some of the possibilities.

Published work on atomic absorption analysis has been reviewed
in [Reference 4]. A variety of techniques has been developed. As
primary sources, continuous radiation from pulsed flash lamps as well
as line radiation from pulsed hollow cathode lamps were used. Free
running and Q-switched solid state lasers, continuously emitting Ar^+
and pulsed CO_2 TEA lasers served as atomization devices. It is
claimed that higher power of detection can be obtained by confining
the vapor in a heated graphite cuvette. In two-step procedures, the
final atomization is accomplished in a separate step which may sub-
stantially improve the power of detection. Of more recent
work[10,16,33,43,44,46,47] it should be mentioned that Petukh et
al.,[16] used a spark cross discharge similar to that in emission
work for the final atomization; Quentmeier et al.,[48,49] an acousto-
optical Q-switch for long-lasting vaporization and efficient atomization; Sumino et al.,[43] combined laser ablation with electrothermal
atomization. Also Dittrich et al.,[44] and Manabe and Piepmeier[46]
used a dye laser for plume atomization combined with a pulsed HCL.
Quillfeldt[47] compared AAS and AES and found similar analytical
properties.

To sum up, atomic absorption methods are basically single element determination methods and may therefore be of limited application possibly in cases where not enough material for several sequential determinations is available. Their limited dynamic concen-

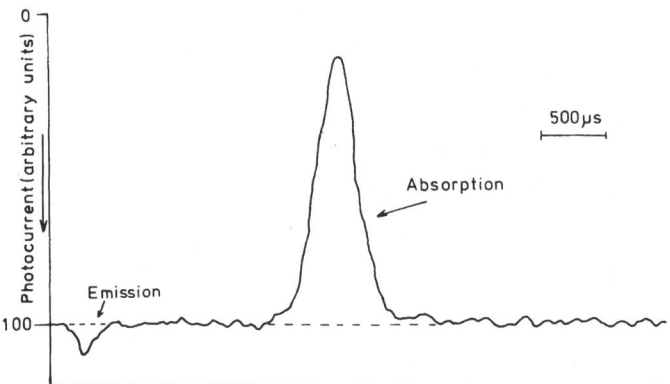

Fig. 4. Time dependence of the photocurrent during laser evaporation
(Height above sample surface, 25 mm).

tration range may hamper the analysis of random samples of completely
unknown composition.

ATOMIC FLUORESCENCE SPECTROSCOPY USING LASER ATOMIZERS

These methods are closely related to AAS methods except that,
due to the emission of radiation, a large concentration range is
available from a single spectral line. From this point of view such
methods are suitable for simultaneous multielement determination
(provided, simultaneous excitation can be effected). So far, only
one application of AFS to laser-produced plume has been reported.
Measures and Kwong[52,58] with their 'Tablaser' atomized sample
material with the help of Q-switched ruby radiation which was then
excited at low pressure to resonance fluorescence by absorption of
radiation from a nitrogen-laser pumped tuned dye-laser. At a spatial
resolution of about 100 µm, limits of detection of the order of 1
µg/g or 10^{-13} g respectively are claimed with the added advantage of
relative freedom from matrix effects.

ATOMIC EMISSION SPECTROSCOPY USING LASER ATOMIZERS

The spectral character of optical emission spectra produced
directly by laser atomizers differs from that of customary arc or
spark discharges. It strongly depends on the power and mode of
operation of the laser. Giant pulses produce a strong background and
broad lines with, however, little or no self-reversal. On the other
hand, unswitched operation with slow switches lead to spectra of
lower continuous background and lines not quite as broad as in the
case of resonance lines with pronounced self-reversal. As the latter
are very often recommended as analytical lines, due care should be

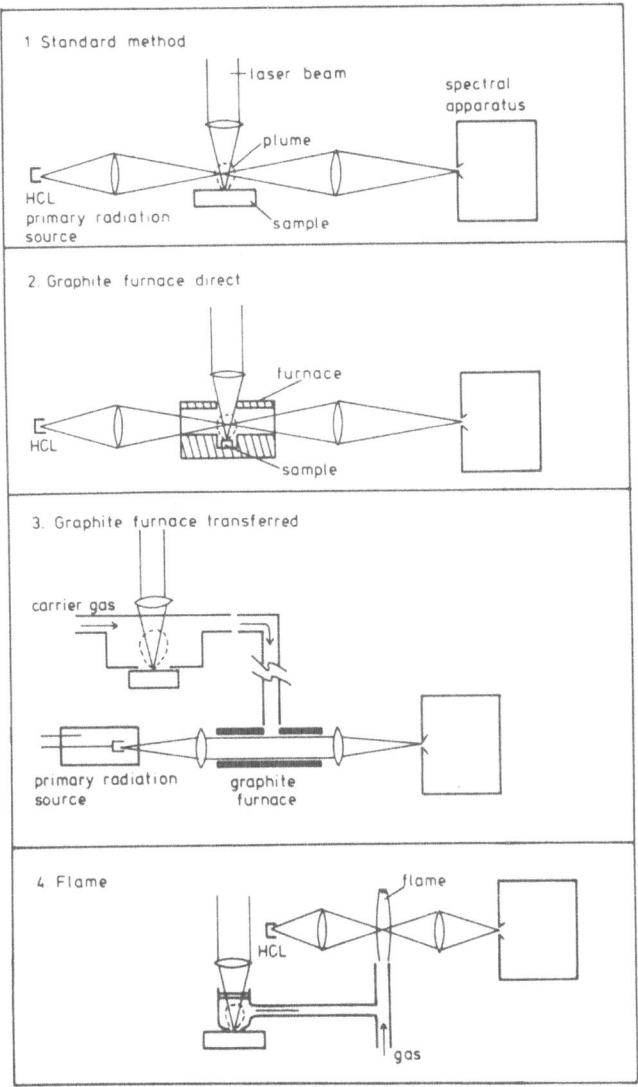

Fig. 5.   Atomic absorption spectroscopy of laser vaporized samples.

exercised by their use. Generally, the quality of the spectra is not
as good as that of other spectrochemical light sources. In Figure 6
some combinations for laser-optical emission spectroscopy are demonstrated.

Substances with a sufficiently high melting point are suited for
direct analysis with the help of the spectrum as generated by the
laser radiation impact. For some kinds of operation, especially if

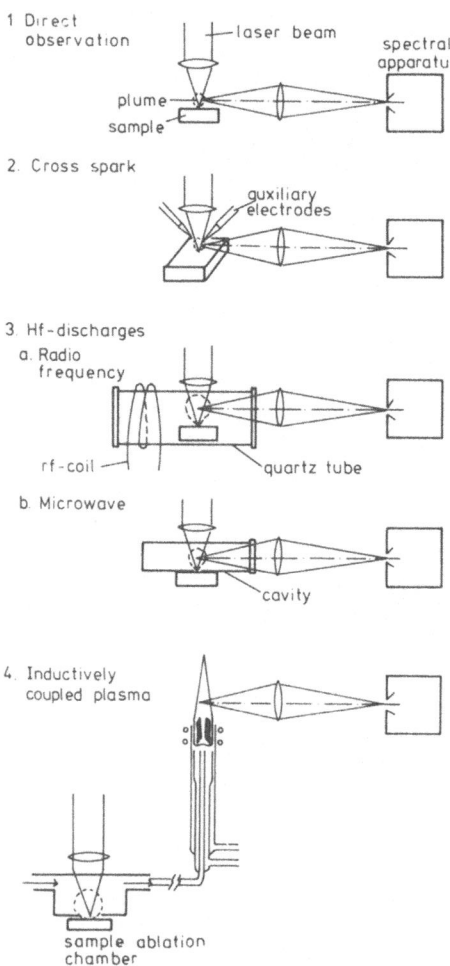

Fig. 6. Possibilities of optical emission spectroscopy with laser atomizers.

high spatial resolution and consequently, low power operation of the laser is necessary, a two-step process is the method of choice, i.e. the laser-produced vapor plume is excited in an additional source. In this configuration, the primary radiation emerging from the vapor cloud is weak as compared to the one produced by the additional excitation which now mainly determines the spectral character of the radiation. Mode of operation of the laser, power and energy are again responsible for the crater dimensions, but of special importance are now also the transport of material into the excitation zone, the time of residence of the laser-produced vapor in this additional source and its thermodynamic properties.

The oldest and most popular method of additional excitation makes use of a spark discharge. This discharge takes place between two pointed electrodes, usually of pure graphite, and is fed from a charged condenser bank via suitable impedances (resistors and inductors). The energy stored in the condenser may be in the order of a joule, the voltage lower than the dielectric breakdown voltage of the electrode gap. The discharge is then initiated by charged particles contained in the vapor cloud entering the gap. The cross discharge should only be effective during the residence time of the laser-produced vapor in the gap. It has been shown that a free choice of the onset of the discharge may give better analytical results than automatic triggering. The cross discharge source closely resembles a customary spectrochemical medium voltage spark source. Therefore, the resulting spectra are also very similar with the result that the spectral lines are much narrower and the spectral background lower as compared to those in the primary spectrum of the laser-produced vapor plume. The intensity ratio of line-to-background is improved by orders of magnitude in favorable cases, and consequently, the power of detection. The gross intensity of the emerging radiation may also be higher by up to two orders of magnitude. As a disadvantage it may be stated that impurities in the auxiliary electrodes cannot be distinguished by their signals from those in the analytical sample. To overcome this ambiguity, electrodeless discharges have been recommended. The vapor can either be produced directly in the excitation zone of an induction coupled plasma (ICP) or a microwave induced plasma (MIP) or it can be transferred into it with the help of a carrier gas. All these methods are treated in detail in[3,4,5]. Some recent results shall be briefly mentioned. Dimitrov et al.,[8] investigated the influence of gas mixtures on the limit of detection with spark cross excitation. They obtained the best results when using pure Argon. In another paper, Dimitrov and Maximova[9] improved the power of detection by inclining the sample surface in respect to the direction of the laser radiation, probably causing a decrease of absorption of the laser radiation in the vapor cloud. Nickel et al.,[53,54] in an extensive study, optimized conditions for quantitative local analysis of graphite with emphasis on good time matching of evaporation and cross excitation. Uchida et al.,[32] concentrated element traces on filter paper after solvent extraction and subjected them to laser vaporization with cross excitation in a spark discharge. Leis and Laqua[50,51] used additional excitation in an MIP, the sample being directly part of the microwave cavity. Limits of detection in the µg/g range were obtained. Ishizuka and Uwamino[42] introduced a laser-produced aerosol into a MIP, obtaining limits of detection ranging from 1 µg/g Zn in Al to 22 µg/g Mn in steel with relative standard deviations from 0.01 to 0.13. Bachurina et al.,[22] transferred the laser-produced aerosol to a CMP (capacitively coupled microwave plasma) for the determination of the alkaline and earth alkaline elements and iron in surface impurities of silicon sheets in the nanogram range. Thompson et al.,[56,57] used the combination of laser and ICP and compared the results with standard ICP analysis of the dissolved steel samples. Generally, with laser, absolute detection limits were lower as compared to the nebulization technique, whereas with concentrational limits of detection it was the other way round.

LASER ATOMIZERS IN MASS SPECTROMETRY

The ions formed in the vapor cloud as a result of the action of focused laser radiation on a solid surface can be separated and detected with a mass spectrometer. In this way, combination of a laser with a mass spectrometer permits the construction of instruments for bulk as well as local or microprobe analysis. Both, double-focusing and time-of-flight mass spectrometers have been used. Earlier work has been fully covered by Conzemius and Capellen[62] in their review article. The modern state of the art regarding bulk analysis, microanalysis and thin film analysis of moderate spatial resolution (crater diameters about 20 μm, crater depths ranging from 0.1 to 10 μm) is represented by the work of Jansen and Witmer[63,64]. They used a Q-switched Nd:YAG laser system as an ion source, Figure 7, and an AEI MS 702 double-focusing mass spectrometer with photographic detection. This permits detection of ions in the mass range between 6 and 240 AMU with good resolution, irrespective of the considerable energy spread of the laser-produced ions. An absolute detection limit of about 10^{-12} g, a relative detection limit of 10 ng/g for bulk, and 1 mg/g for micro-analysis can be obtained. The precision can be described by a relative standard deviation of about 0.2, the accuracy by a factor of 2 without and 0.2 with standards.

A completely different line of development has been pursued by Hillenkamp et al.,[66] resulting in a laser microprobe mass analyzer (LAMMA) of which a commercial model, the LAMMA-500, (Leybold-Heraeus) is available. Figure 8 shows the principles of this instrument, in which a Nd:YAG laser irradiated the sample which may be a foil or a thin layer on a transparent substrate. The ions formed enter a time-of-flight mass spectrometer for separation and detection by an open Cu-Be secondary electron multiplier. The system is preferentially used for the analysis of thin samples, mostly biological material. Absolute detection limits lie in the range of 10^{-18} to 10^{-20} g. Under favorable circumstances (very thin films) a lateral resolution of less than 1 μm is obtainable. Not only atoms, but molecules as well can be detected.

A similar instrument has been described by Dingle and Griffith[71] which in addition, enables the analysis of bulk samples by means of reflected ions. A compensation for the initial spread of energy of the ions is also provided.

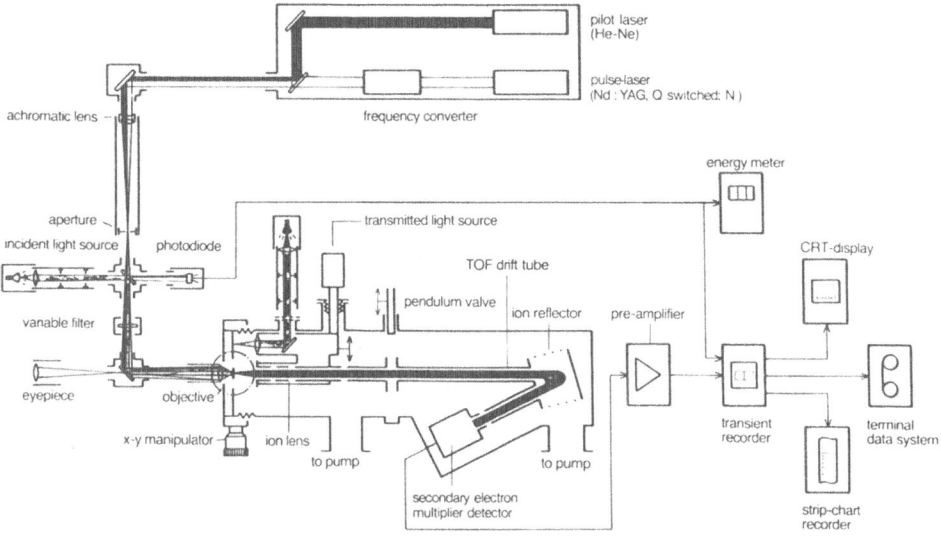

Fig. 7.   Schematic diagram of LAMMA 500.

LASER SPECTROCHEMICAL PROCEDURES

Local Analysis

A microanalysis performed in situ may be called local analysis.
Typical examples are an inclusion in a metal or mineral or the changing composition along a welding seam. In such cases, spatial resolution and the concentration range of the elements to be determined
influence significantly the choice of laser operating parameters,
spectral apparatus and radiation detector. For a one-shot analysis
with no additional excitation, Q-switched lasers producing craters of
about 50 µm diameter may yield line intensities which are sufficiently strong to be recorded with a high speed spectrograph (f/5
or better) if only the main constituents of a sample have to be
determined. With additional excitation and free running lasers,
usable analytical signals can be obtained from craters with a diameter of about 10 µm. In this case, a medium speed spectrograph may
be sufficient. Photoelectric detection is much more sensitive but
does not easily permit the recording of a substantial part of the
spectrum. An interesting attempt has been made by Talmi et al.,[39]
to overcome these difficulties. They used SIT-vidicons and second
generation photodiode arrays as detectors and obtained limits of
detection between 2 and 500 µg/g in surface analysis and depth pro-
filing. Unfortunately, present day vidicons and silicon photodiode
arrays cannot optimally be matched to commercial spectral apparatus.
In Table 2 the properties of laser local analysis are summarized.

Table 2. Laser Optical Spectroscopy Pertaining to Local Analysis

Property	Laser
Spatial resolution	
Diameter of area, μm	10-25 (depending on sample);
Depth, μm	widely variable, depending on working conditions, > 3;
Determinable elements	~70; elements with persistent lines in VUV (halogens, gases, etc. not yet investigated);
Multi-element determination	simultaneous; sequential restricted because destructive;
Sample preparation	no;
Atmosphere	any, (vacuum not recommended);
Destructive analysis?	yes;
Concentration range	.1 mg/g up to 1 g/g, lower limit depending on material and spatial resolution,
Quantitative analysis	yes, provided probed micro volume in itself is homogeneous;
Determination of elemental distribution in the sample	yes, but limited by coarse resolution and slow speed;
Precision	medium;
Speed	slow with photographic, fast with photoelectric methods;
Cost	low to medium.

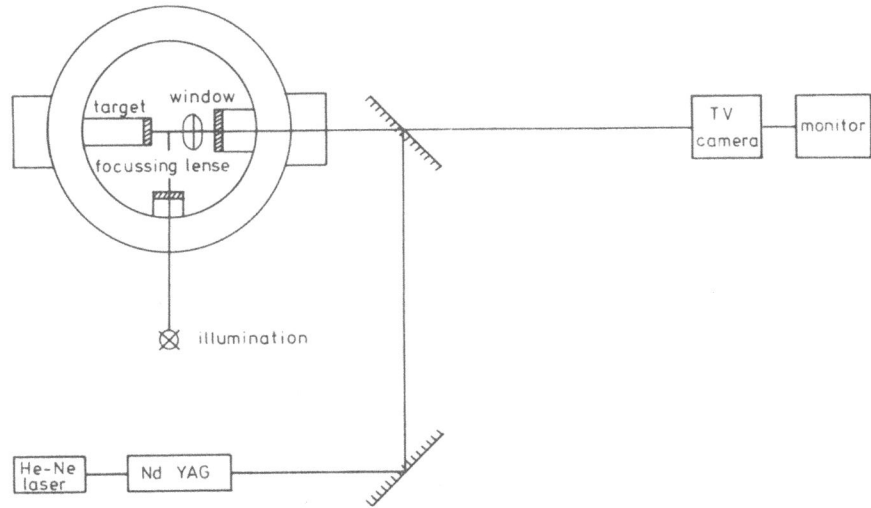

Fig. 8.   Schematic structure of laser ion source system. [From
          J. A. J. Jansen et al.,64].

Surface Analysis and Depth Profiling

Surface analysis represents local analysis in which the depth of
erosion is kept as small as possible. Either a free running laser of
very low output energy and additional excitation or a Q-switched
low-energy, high-power giant pulse should be used. The former may
yield crater diameters of 10 μm with a depth of 3 μm, the latter
diameters of about 25 μm and a depth of 1 μm depending also on the
sample material.

Microanalysis

Techniques for microanalysis are similar to those for local
analysis. A microsample must be transformed into a small volume and
then atomized for local analysis. For minimum matrix effects, the
sample should be vaporized completely. The amount of sample can be
as low as 1 μg. Limits of detection down to 10 μg/g can be achieved,
depending on the element and the size of the sample.

Macro- or Bulk Analysis

Because of the unique property of laser atomization, homogeneous, non-conducting solid materials in addition to electrically
conducting materials can be directly analysed by means of optical
spectroscopy. There is some reason to recommend laser atomization

for electrically conducting materials normally analysed with the help
of spark discharges. In electrical discharges, the burning spot
wanders erratically across the surface which may give rise to systematic errors if that movement is influenced by a structure of the
surface. Attempts have therefore been undertaken to prevent the
wandering of the burning spot by means of focused laser radiation[19]
resulting in improved precision and accuracy.

In laser analysis of homogeneous samples, precision can be
improved by increasing the number of shots per analysis. This should
be done by automatic scanning of the surface in such a way that by
each laser shot, a hitherto unattacked part of the surface is
sampled. The basic obtainable precision depends on the mode of
operation. It can be as good as 0.1 for a single shot without additional excitation and correspondingly less for a multishot analysis. If high power of detection is required and hence, additional
excitation, the basic precision may degrade to about 0.3. These
figures apply to laser output energies of about 1 joule which corresponds to vaporized sample material of about 10 µg per shot.
Sampling of much more material for a single-shot analysis with laser
output energies of up to 100 joule has been reported. A good compro-
mise would be a multishot analysis with energies of about 5 joule per
shot. In macro-analysis, non-compromise conditions can be selected
for good power of detection. In Table 3, some typical examples are
given.

Special Procedures

Artificial gas atmospheres and pressures other than ambient can
be used to further optimize analytical conditions. As already mentioned and confirmed by other investigators, argon seems to be favorable for high intensity of the spectral lines and reduced background.
For the best line-to-background ratio vacuum was established as
optimum.

The laser cloud is inhomogeneous and very transient in time and
space. By time-resolving and gating of the analytical signal conditions for optimum line-to-background ratio can be established.
Space-resolved spectroscopy can be successfully applied whenever the
analysis is performed without auxiliary excitation due to the fact
that spectral lines and spectral background have their maximum intensity in different zones of the laser plume. Finally, a magnetic
field may also be used to improve line-to-background ratios and
consequently, the power of detection. The best results in taking
account of many single step improvements have been reported by Treytl
et. al.,[73]. Table 4 lists some of their results. It should
further be noted that this group, in favorable cases, was able to
secure from biological materials, useful analytical signals from
craters of about 1 µm diameter.

Table 3. Limits of Detection (µg/g) for some Elements in Optical
          Emission Spectrochemical Analysis pertaining to several
          Excitation Methods

Analytical line, nm*	Laser alone*	Cross-excitation*	85MHz ICP*	2,47GHz MIP**	Analytical line, nm**
Mg 285.2	10	0.9	0.2	0.4	Mg 279.6
"	20	6	4	2	"
"	–	1	–	–	"
"	–	6	3	2	"
"	20	1	1	–	"

* [Values taken from Möde,72].
**[Values taken from Leis et al.,51].

Table 4. Use of Time Resolution for Obtaining Low Limits of
          Detection in Analytical Procedure for Analysis of Biologic
          Materials, Employing Q-switched Single Laser Pulse and
          Electronic Gating Technique*.

Element	Wavelength, nm	Delay time, µs	Integration time, µs	Detection limit, pg
Li	610.4	4	5	0.2
Mg	279.6	4	7	0.002
Ca	393.4	5	15	0.01
Fe	302.0	10	6	0.3
Cu	324.8	15	3	0.002
Zn	213.9	5	2	0.05
Hg	253.7	7	3	0.3
Pb	405.8	16	5	0.1

*[From Treytl et al.,73].

In an interesting application, Adrian et al.,[29] used a computerized laser-micro-spectrograph equipped with a SIT-vidicon for
remote analysis at a nuclear reactor. Radiation from the laser head
with additional excitation was conducted to the spectral apparatus by
means of an optic fiber of 40 mm length.

CONCLUSION

    Atomization by means of focused laser radiation has supplemented
the analytical repertoire considerably due to its applicability to

all kinds of samples, irrespective of their chemical and physical
conditions. Originally introduced for local and micro-analysis by
optical emission spectroscopy, it is now also used for bulk and
surface analysis of a large variety of materials, in combination with
other spectroscopic methods like AAS and mass spectrometry. Whereas
upt to now, mostly qualitative and semiquantitative procedures have
been described, there are high precision procedures available waiting
for their introduction into analytical practice. Reliable calibration in local analysis remains a problem which needs further attention. Laser sources for atomization have been developed to a high
degree of perfection and can be operated reliably and at low costs.
Excimer lasers may find special uses resulting from their high power
output and high repetition rates.

REFERENCES

1. T. H. Maiman, Stimulated optical radiation in ruby, Nature, 187:
    493 (1960).
2. F. Brech and L. Cross, Optical micro-emission stimulated by a
    ruby MASER, Appl.Spectrosc., 16:59 (1962).
3. R. H. Scott and A. Strasheim, Laser emission excitation and
    spectroscopy, in: "Applied Atomic Spectroscopy", Vol.1, E. L.
    Grove, ed., Plenum Press, New York and London (1978).
4. K. Laqua, Analytical spectroscopy using laser atomizers, in:
    "Analytical Laser Spectroscopy", N. Omenetto, ed., John Wiley
    & Sons, New York, Chichester, Brisbane, Toronto (1979).
5. M. L. Petukh and A. A. Yankowskii, Atomic emission spectral
    analysis using lasers, Zh.Prikl.Spektrosk., 29:1109 (1978).
6. H. Moneke and L. Moenke-Blankenburg, "Laser Micro-Spectrochemi-
    cal Analysis", Adam Hilger, London (1973).
7. W. van Deijck, J. Balke, and F. J. M. J. Maessen, An assessment
    of the laser microprobe analyser as a tool for quantitative
    analysis in AES, Spectrochim.Acta, 35B:359 (1979).
8. G. Dimitrov, L. Nikolova, and Ya. Vasilev, Influence of the dis-
    charge gas on the emission and thermal characteristics of
    laser-induced microplasmas, Mikrochim.Acta, 1:503 (1979).
9. G. Dimitrov and Ts. Maximova, Improvement of the reproducibility
    and sensitivity of laser microspectral analysis, Spectroscopy
    Letters, 14:737 (1981).
10. E. K. Vul'fson, V. I. Dvorkin, and A. V. Karyakin, The problem
    of Vaporization of a substance in a laser jet, Zh.Prikl.
    Spektrosk., 32:414 (1980).
11. V. A. Ageev, A. V. Kolesnik, and A. A. Yankovskii, Laser deter-
    mination of electroplating thickness, Zh.Prikl.Spektrosk.,
    26:360 (1977).
12. A. F. Bokhonov, V. S. Burakov, V. V. Zhukovskii, and A. A.
    Stavrov, Erosion due to radiation activity of lasers in the
    mode-locked regime, Zh.Prikl.Spektrosk., 26:821 (1977).

13. S. I. Anisimov, Ya. A. Imas, G. S. Romanov, and Yu. V. Khodyko, "The Effect of High Power Radiation on Metals", (in Russian), Nauka, Moscow (1970).

14. M. F. Stel'makh, "Lasers in Technology", (in Russian), Energiya, Moscow (1975).

15. S. P. Atamanova, LMA-1 apparatus for the microspectral analysis of certain minerals found in the Kolsk Peninsula, Zh.Prikl. Spektrosk., 32:202 (1980).

16. M. L. Petukh, A. D. Shirokanov, and A. A. Yankovskii, Use of laser pulses with electric discharges for atomic absorption analysis, Zh.Prikl.Spektrosk., 32:414 (1980).

17. A. A. Yankovskii, "Quantum Electronics and Laser Spectroscopy", (in Russian), Nauka i Tekhnika, Minsk, (1979).

18. A. N. Zaidel, G. V. Ostrovskaya, and Yu. I. Ostrovskii, "Technique and Application of Spectroscopy", (in Russian), Nauka, Moscow (1972).

19. V. A. Ageev, A. V. Kolesnik, and A. A. Yankovskii, Possibility of limiting the migration of current-carrying discharge channels by a laser bean, Zh.Prikl.Spektrosk., 26:417 (1977).

20. W. Maul and W. Quillfeldt, Homogeneity investigation with the LMA 10 laser microspectral analyser, Jena Review 22:234 (1977).

21. E. Litz, Staubanalysen mit dem laser-mikrospektral-analysator LMA 10, Jenaer Rundschau, 5:237 (1977).

22. L. G. Bachurina, V. M. Perminova, and S. A. Savostin, Spectral micro-analysis with laser and plasma excitation of the spectra, Zavodsk.Lab., 45:1113 (1979).

23. D. E. Maksimov, N. K. Rudnevskii, V. P. Ryabchikova, and E. N. Pryanichnikova, Laser spectral micro-analysis of welded seams, Zavodsk.Lab., 45:333 (1979).

24. D. E. Maksimov, N. K. Rudnevskii, V. P. Ryabchikova, S. M. Chekhonin, I. V. Shlyapnikov, and I. S. Shklyaeva, LMA-1 laser micro-analyzer applied to welds in alloy steels, Zavodsk.Lab., 43:445 (1977).

25. M. B. Kozik, Modification of a quantitative laser-spectrographic method of determination of cations contained in tissue slices, Folia Histochem.Cytochem., 17:153 (1979).

26. W. Klimecki, Local spectral analysis and lasers, Chem.Anal. (Warsaw), 23:3 (1978).

27. R. M. Manabe, Effects of atmospheric pressure on line widths and spatial distributions of transient AA signals of minor constituents in metal samples atomized by a dye laser microprobe, Dissertation, Oregon State University (1977).

28. M. B. Kozik, Laser-spectrographic study on the contents of metals in brains of patients with arteriosclerotic dementia, Folia Histochem.Cytochem., 16:31 (1978).

29. R. S. Adrain, R. C. Klewe, and E. J. Ormerod, Robust portable computerized laser-microspectrograph, in: Conf. Proceed. "Electro-Optics Laser International '80", ICP Science and Technology Press, Ltd.

30. R. S. Adrain, D. R. Airey, R. C. Klewe, and E. J. Ormerod,
     Trace element line intensities in laser produced metal vapor
     plasmas, Private communications.

31. R. S. Adrain, J. Watson, P. H. Richards, and A. Maitland, Laser
     microspectral analysis of steels, Opt.Laser Technol., 12:
     137 (1980).

32. H. Uchida and K. Iwasaki, Laser-microprobe spectroscopy for
     trace elements concentrated on filter-paper after solvent
     extraction, Bunseki Kagaku, 25:752 (1976).

33. T. Ishizuka, Y. Uwamino, and H. Sunahara, Laser-vaporized
     atomic absorption spectrometry of solid samples, Anal.Chem.,
     49:1339 (1977).

34. J. Kozak, Correction for the volume of vaporized material in
     laser spectral analysis, Chem.Listy., 71:424 (1977).

35. M. Hufner, Using the KSR 4200 small process-control computer
     with laser emission spectrochemical analysis, Jena Review,
     21:312 (1976).

36. J. M. Green, W. T. Silfvast, and O. R. Wood, Evolution of a CO_2-
     laser produced Cadmium plasma, J.Appl.Phys., 48:2753 (1977).

37. G. Dimitrov and I. Koleva, Investigations of sparking with the
     aid of a LMA-1 laser microspectral analyser, Chem.Anal.
     (Warsaw), 22:861 (1977).

38. E. Raitieri, M. Guerzoni, and G. Grammatica, Use of microprobe
     laser in emission spectrography, Met.Ital., 73:173 (1981).

39. Y. Talmi, H. P. Sieper, and L. Moenke-Blankenburg, Laser micro-
     probe elemental determinations with an optical multichannel
     detection system, Anal.Chim.Acta, 127:71 (1981).

40. A. V. Karyakin, E. K. Vul'fson, and A. F. Yanushkevich, "Über
     die Möglickeiten und Grenzen der Anwendung des Laser Atomis-
     ators für die Analyse fester Proben", Publ. Technische
     Hochschule Karl-Marx-Stadt (1976).

41. J. Mohr, Probenkammer mit versorgungseinheit zum laser mikro-
     spektral analysator LMA 10, Jenaer Rundschau, 24:245 (1979).

42. T. Ishizuka and Y. Uwamino, Atomic emission spectrometry of
     solid samples with laser vaporization-microwave-induced
     plasma system, Anal.Chem., 52:125 (1980).

43. K. Sumino, R. Yamamoto, F. Hatayama, S. Kitamura, and H. Itoh,
     Laser atomic absorption spectrometry for histochemistry,
     Anal.Chem., 52:1064 (1980).

44. K. Dittrich and R. Wennrich, AAS using laser vaporization fol-
     lowed by ETA, Spectrochim.Acta, 35B:731 (1980).

45. E. K. Wulfson, W. I. Dworkin, and A. W. Karyakin, Measurements
     on the thermal conditions in the vapor cloud produced by
     laser impact on graphite, Spectrochim.Acta., 35B:11 (1980).

46. R. M. Manabe and E. H. Piepmeier, Time and spatially resolved
     atomic absorption measurements with a dye-laser plume
     atomizer and pulsed hallow-cathode lamps, Anal.Chem., 51:
     2066 (1979).

47. W. Quillfeldt, Combination of AAS and AES with the laser
     spectrometer LMA 10, Jena Review, 23:226 (1978).

48. A. Quentmeier, K. Laqua, and W. -D. Hagenah, Atomic-absorption spectrometry by laser-radiation evaporation of solid samples. I. Optimization of experimental parameters, Spectrochim.Acta, 34B:117 (1979).

49. A. Quentmeier, K. Laqua, and W. -D. Hagenah, Atomic absorption spectroscopy of solid samples using evaporation by laser radiation. II. Analytical applications, Spectrochim.Acta, 35B:139 (1980).

50. F. Leis and K. Laqua, Emission spectrometric analysis using microwave excitation of a vapor cloud produced by laser impact on a solid. I. Principles of the method and experimental realization, Spectrochim.Acta, 33B:727 (1978).

51. F. Leis and K. Laqua, Emission spectrometric analysis using microwave excitation of a vapor cloud produced by laser impact on a solid. II. Analytical applications, Spectrochim. Acta, 34B:307 (1979).

52. R. M. Measures and H. S. Kwong, Tablaser: trace (element) analyser based on laser ablation and selectively excited radiation, Appl.Opt., 18:281 (1979).

53. H. Nickel, F. A. Peuser, and M. Mazurkiewicz, Evaporation of material and influence of auxiliary spark gap on the spectral excitation by means of laser emission spectroscopy for local analysis of graphite, Spectrochim.Acta, 33B:675 (1978).

54. H. Nickel, F. A. Peuser, M. Mazurkiewicz, and W. Dörge, Quantitative mikroanalyse von reaktorgrafit mit hilfe der laser-emissions-spektroskopie, Jenaer Rundschau, 5:199 (1979).

55. S. O. Baisane, V. S. Chincholkar, and B. N. Maltor, Laser mikro-spektralanalyse einige forensische applikationen, Jenaer Rundschau, 5:206 (1979).

56. M. Thompson, J. G. Goulter, and F. Sieper, Verdampfung fester proben im laserstrahl des LMA 10 und ihre zuführung in induktiv-gekoppeltes plasma (ICP) für die atom-emissions-spektrometrie, Jenaer Rundschau, 5:202 (1981).

57. M. Thompson, J. G. Goulter, and F. Sieper, Laser ablation for the introduction of solid samples into an inductively coupled plasma for atomic-emission spectrometry, Analyst., 106:32 (1981).

58. R. M. Measures and H. S. Kwong, Trace Element laser microanalyser with freedom from chemical matrix effect, Anal. Chem., 51:428 (1979).

59. D. C. Smith, Laser radiation-induced air breakdown and plasma shielding, Opt.Engineering, 20:962 (1981).

60. Kh. I. Zil'bershtein, Modern light sources for analysis of optical emission spectra, Zavodsk.Lab., 46:1095 (1980).

61. R. E. Honig and J. R. Woolston, Laser-induced emission of electrons, ions and neutral atoms from solid surfaces, Appl. Phys.Lett., 2:138 (1963).

62. R. J. Conzemius and J. M. Capellen, A review of the application to solids of the laser ion source in mass spectrometry, Int. J.Mass Spectrom.Ion Phys., 34:197 (1980).

63. J. A. J. Jansen and A. W. Witmer, Spark source mass-spectrometry in the research laboratories of an electronic industry, Fresenius Z.Anal.Chem., 309:262 (1981).

64. J. A. J. Jansen and A. W. Witmer, Quantitative inorganic analysis by Q-switched laser mass spectroscopy, Spectrochim.Acta, 37B:483 (1982).

65. E. Denoyer, R. van Grieken, F. Adams, and D. F. S. Natusch, Laser microprobe mass spectrometry. I. Basic principles and performance characteristics, Anal.Chem., 54:26A (1982).

66. F. Hillenkamp, E. Unsöld, R. Kauffmann, and R. Nitsche, Laser microprobe mass analysis of organic materials, Nature, 256: 119 (1975).

67. M. H. R. Hutchinson, Excimer and Excimer Lasers, Appl.Phys., 21:15 (1980).

68. P. Boissel, G. Hauchecorne, F. Kerteve, and G. Mayer, Optical elements using oscillating gazes, Opt.Communications, 4:44 (1971).

69. J. F. Ready, "Effects of High-Power Laser Radiation", Academic Press, New York and London, (1971).

70. V. G. Mossotti, K. Laqua, and W. -D. Hagenah, Laser-microanalysis by atomic absorption, Spectrochim.Acta, 23B:197 (1967).

71. T. Dingle and B. Griffith, A laser ion mass analyser (LIMA) for bulk samples with high spatial resolution and PPM hydrogen sensitivity, J.Phys.E, Scient.Instrum., 14:513 (1981).

72. U. Möde, Extension and enhancement of the spectral emission of radiation from samples vaporized by laser radiation, Ph.D. Thesis, Münster (1970).

73. W. J. Treytl, J. B. Orenberg, K. W. Marich, A. J. Saffir, and D. Glick, Detection limits in analysis of metals in biological materials by laser microprobe optical emission spectrometry, Anal.Chem., 44:1903 (1972).

THEORY OF RESONANCE IONIZATION SPECTROSCOPY*

M. G. Payne and G. S. Hurst

Oak Ridge National Laboratory
Oak Ridge
Tennessee, USA

INTRODUCTION

Resonance Ionization Spectroscopy (RIS) can be defined as a state selective detection process in which pulsed tunable lasers are used to promote transitions from the selected state of the atoms or molecules in question to higher states, one of which will be ionized by the absorption of another photon. At least one resonance step is used in the stepwise ionization process, and it has been shown[1-4] that the ionization probability of the spectroscopically selected species can nearly always be made close to unity. Since measurements of the number of photoelectrons or ions can be made very precisely and even one electron (or under vacuum conditions, one ion) can be detected, the technique can be used to make quantitative measurements of very small populations of the state-selected species. Counting of individual atoms has special meaning for detection of rare events.

The ability to make saturated RIS measurements opens up a wide variety of applications to both basic and applied research. In reviews of RIS[1-4,5] the subject was treated generally, including the underlying photophysics applications, the ability to use it to count single atoms, and its applications to measurements in atomic and molecular physics. We view resonance ionization spectroscopy as a specific type of multiphoton ionization in which the goal is to make quantitative measurements of quantum-selected populations in atomic or molecular systems. This goal is attained by requiring that

*Research sponsored by the Office of Health and Environmental Research, U.S. Department of Energy under contract W-7405-eng-26 with the Union Carbide Corporation.

the selective excitation steps be resonant in nature and involve only one- or two-photon (only one-photon if at all possible) absorption processes, thereby allowing the entire process to be carried to saturation without loss of spectroscopic selectivity due to laser power induced shifts or broadening.

THE FUNDAMENTAL RIS SCHEME

It now appears that all atoms except helium and neon can be efficiently ionized by using a scheme in which a sequence of single-photon absorptions promote a valence electron to a high-lying excited state with $\ell \geq 1$, from which ionization occurs due to a relatively intense beam of Nd-YAG radiation at 1.06 μm or CO_2 radiation at ~10 μm. We will now discuss the laser requirements for a simplified version of an RIS scheme involving only one-photon absorption.

It is well known[1-4,5,6,7] that when Ω_R (the Rabi frequency) and Δ_S (the a.c. Stark shift) for a resonant transition are small compared with the laser bandwidth, a rate equation analysis can be used in describing the interaction of an atom with the laser fields. In particular, in the absence of collisions, which redistribute atomic populations among the magnetic substates within the laser pulse length, we find for copropagating, plane polarized beams that the cross sections for absorption and stimulated emission are equal and are given by

$$\sigma_a = \sigma_s$$
$$= \frac{2\pi^2}{\sqrt{2\pi\sigma^2}} \left(\frac{e^2}{M_e c} \right) F_{01} \, e^{-\frac{\delta^2}{2\sigma^2}} \tag{1}$$

where $\delta = \overline{\omega} - \omega_r$ is the detuning of the line center of the laser from the resonant angular frequency for the transition, ω_r; F_{01} is the absorption oscillator strength of the transition; $2\sqrt{2\ln 2}\, \sigma$ is the full width at half maximum (FWHM) of the laser line shape in angular frequency units; and the laser spectrum is taken to be $(2\pi\sigma^2)^{-1/2} \exp[-(\omega - \overline{\omega})^2/2\sigma^2]$. Thus,

$$\sigma_a(cm^2) = 8.32 \times 10^{-13} (F_{01}/FWHM(cm^{-1})), \tag{2}$$

where FWHM (cm^{-1}) is the laser bandwidth in cm^{-1}. Thus, for a transition with $F_{01} = 0.25$ and a laser bandwidth FWHM $(cm^{-1}) \approx 0.2\ cm^{-1}$, we find $\sigma_a \approx 1 \times 10^{-12}\ cm^2$.

For most atoms it is most difficult to use one-photon absorption for transitions from the ground state to one of the lower lying electronically excited states. In particular, the rare gases, most of the halogens, hydrogen, oxygen, and carbon all require vacuum ultraviolet (VUV) radiation for this initial step. Once the initial excitation is made, other transitions can be pumped by a relatively weak dye laser, usually without even resorting to frequency doubling or mixing. For this reason we focus our attention on the situation in Figure 1. Figure 1 a very weak laser field is used to pump the transition $|0>$ to $|1>$. The population of $|1>$ can be promoted to a higher excited state and ionized on a time scale that is very short compared with the laser pulse length by using only dye lasers with

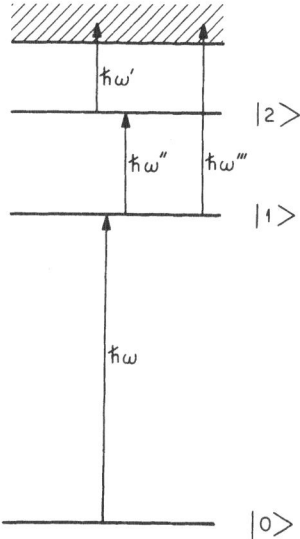

Fig. 1. Energy level diagram for a resonance ionization process involving only one-photon stepwise processes.

FWHM (cm^{-1}) $\approx$ 0.2 cm^{-1}, power density = I < 2kw/cm^2 and pulse length $\leq$ 5 x 10^{-9} sec. The Nd-YAG light used for ionization may have I $\approx$ 5 x 10$'$ W/cm^2 (i.e., 50 mJ pulses of beam area 0.08 cm^2 and pulse length ~10^{-8} sec). The ionizing light will be close to ionization threshold, and in ionizing p, d, or f states the ionization cross section will nearly always exceed 10^{-17} cm^2. The ionization of excited s states should be avoided in all but a few light elements due to Cooper minima relatively near threshold and their general characteristic of having ionization cross sections that are two or three orders of magnitude smaller than those of p, d, or f states in the region near threshold.

Since it is easy to deplete $|1>$ by ionization on a time scale that is very fast, the repopulation of $|1>$ due to stimulated or spontaneous emission is negligible when the $|0> -> |1>$ transition is pumped at a much slower rate. Thus, if $\rho_{00}(\rho,t)$ is the probability that an atom located at a distance ρ from beam axis is in the ground state at time t and F(ρ,t) is the photon flux at ρ at time t:

$$\frac{d\rho_{00}}{dt} (\rho,t) = - \sigma_a F(\rho,t)\rho_{00}(\rho,t). \tag{3}$$

Thus, $P_I(\rho)$, the probability of an atom at ρ being ionized, is given by ($|1\rangle$ is depleted very rapidly)

$$\begin{aligned} P_I(\rho) &= 1-\rho_{00}(\rho,_{00}), \\ &= 1-\exp\left[-\sigma_a \int_0^\infty F(\rho,t)dt\right], \\ &= 1-\exp\left[-\sigma_a\phi(\rho)\right], \end{aligned} \tag{4}$$

where $\phi(\rho)$ is the number of photons/cm^2 at ρ. In order for Equations (3) and (4) to hold for all ρ values where $P_I(\rho)$ is significant, the dye and Nd-YAG lasers used to ionize $|1\rangle$ have beam radii about three times larger than that used for the $|0\rangle$ to $|1\rangle$ transition.

If N is the concentration of the selected population and $\phi(\rho) = \phi(0)\exp(-2\rho^2/d^2)$,

N_I = number of ions produced in length L of the laser beams

$$\begin{aligned} &= NL \int_0^\infty 2\pi\rho\ P_I(\rho)d\rho, \\ &= \frac{N\pi d^2 L}{2} \int_0^\infty x dx\left[1-\exp\left(-\sigma_a\phi(0)e^{-x^2}\right)\right], \\ &= \frac{\pi NLd^2}{2} T\left(\sigma_a\phi(o)\right). \end{aligned} \tag{5}$$

The function $T(x)$ has the power series

$$T(x) = \sum_{n=1}^\infty \frac{(-1)^{n+1}x^n}{n(n!)}. \tag{6}$$

The number of photons is related to d and $\phi(0)$ by $N_p = \pi\phi(0)d^2/2$. Thus, if d = 0.1 cm, $N_p = 2 \times 10^{10}$ photons/pulse, $F_{01} = 0.19$, and FWHM (cm^{-1}) = 0.2 cm^{-1}, we find $\sigma_a\phi(0) \simeq 1$ and $N_I = NL\pi(d^2/2)(0.39) = NL*(6.1 \times 10^{-3}$ cm^2).

FOUR-WAVE MIXING IN XENON AND MERCURY

Important advances in VUV generation have been made by several workers[8-11]. Cotter[12] and Wallenstein et al.,[13] have shown that third-harmonic generation in argon, krypton, and xenon can yield 1-10 W peak power for much of the region 1100 Å $\leq \lambda \leq$ 1470 Å. Four-wave mixing with a two-photon resonant enhancement has been used by

Table 1. Four-Wave Mixing. Table of two- and three-photon resonance transitions in Xe and Hg. Each three-photon resonance borders a region of resonant enhancement for the process $2\omega_1 + \omega_2$ or $2\omega_1 - \omega_2$.

SYSTEM	$|0\rangle$	$|1\rangle$	$|2\rangle$	λ_1	λ_2	$\lambda_r \to 2\omega_1 + \omega_2$
Xe	$5p^6$	$5p^56p$	$5p^57s$	2496Å	1.879μ	1170.3Å
Xe	$5p^6$	$5p^56p$	$5p^58s$	2496Å	9247Å	1099.7Å
Xe	$5p^6$	$5p^56p$	$5p^59s$	2496Å	7506Å	1070.4Å
Xe	$5p^6$	$5p^56p$	$5p^510s$	2496Å	6818Å	1054.9Å
Xe	$5p^6$	$5p^56p$	$5p^511s$	2496Å	6463Å	1046.1Å
Xe	$5p^6$	$5p^56p'$	$5p^55d$	2226Å	10,128Å	1250.2Å
Xe	$5p^6$	$5p^56p'$	$5p^56s$	2226Å	4584Å	1470Å*
Xe	$5p^6$	$5p^56p'$	$5p^56s'$	2226Å	7890Å	1296Å*
Xe	$5p^6$	$5p^56p$	$5p^57d$	2496Å	8327Å	1085Å
Xe	$5p^6$	$5p^56p$	$5p^57s$	2560Å	1.41μ	1170.3Å
Hg	$6s^2$	$6s7s$	$6s6p$	3128.5Å	1.014μ	1849.5Å*
Hg	$6s^2$	$6s7s$	$6s7p$	3128.5Å	1.744μ	1435.5Å
Hg	$6s^2$	$6s7s$	$6s8p$	3128.5Å	7731Å	1301Å
Hg	$6s^2$	$6s7s$	$6s9p$	3128.5Å	6236Å	1250.6Å
Hg	$6s^2$	$6s7s$	$6s10p$	3128.5Å	5805Å	1232.2Å
Hg	$6s^2$	$6s8s$	$6s6p$	2688.0Å	4917Å	1849.5Å*
Kr	$4p^6$	$4p^56p$	$4p^55s$	1934.9Å***	4455Å	1235.8Å*
Kr	$4p^6$	$4p^55p'$	$4p^55s'$	2023.1Å**	7887Å	1164.9Å*

*$2\omega_1 - \omega_2 \to$ WIDE RANGE OF TUNABILITY ON EITHER SIDE OF LINE.
**BY RAMAN SHIFTING 1936Å IN $H_2 \leftrightarrow$ 1936Å BY ArF AMPLIFIER.
***INJECTION LOCKED ArF LASER.

Wallenstein et al.,[13] and Tomkins and Mahon[14] in order to generate several μJ/pulse in mercury. In particular, the latter workers have generated several μJ at 1250.2 Å, and this provides a very effective source to begin the RIS of xenon.

Very crudely, the presence of intense laser light in a gaseous medium causes the atomic wave function to evolve from $|\psi(-\infty)\rangle = |0\rangle$ to $|\psi(t)\rangle = \underset{n}{S} a_n(t)\exp[-i\omega_n t]|n\rangle$.

In particular, if light is present which is two-photon resonant with a higher state, then $a_0(t)$ is reduced but the amplitude for the higher state increases. When intense light is also present which is close to resonance between the two-photon resonance and a higher excited state $|L_3\rangle$ having allowed transitions back to the ground state, $|\psi(t)\rangle$ has three significant amplitudes (i.e., the ground state, the two-photon resonance state, and the near three-photon resonance). Thus, if we consider the expectation value of the electronic dipole operator, $\vec{P}(t) = \langle\psi(t)|\hat{P}|\psi(t)\rangle = \underset{m}{S}\underset{n}{S} a_n^*(t)a_m(t) \exp[i(\omega_n-\omega_m)t] P_{nm}$. The amplitude of $|L_3\rangle$ is enhanced both because of the two-photon resonance and the near three-photon resonance induced by the second laser. Since $|\langle 0|\hat{P}|L_3\rangle| \neq 0$, the individual atoms develop strong dipoles oscillating at $2\omega + \omega$, where ω is the angular frequency of the laser which is tuned near two-photon resonance.

In order to achieve a strong signal a $2\omega_1 + \omega_2$, the pressure (and perhaps a buffer gas partial pressure as well) must be adjusted so that a large degree of constructive interference occurs at the desired wavelength. When the laser beams are unfocused, strong constructive interference occurs when $\Delta k = 2k(\omega_1) + k(\omega_2) - k(2\omega_1 + \omega_2) = 0$, where $k(\omega)$ is the length of the wave vector for light at frequency ω. This is just the condition for the phase of a driven oscillator to be such that it generates a radiation field which is in phase with the combined field due to other atoms upstream in the laser beam.

Table 1 shows levels in xenon and mercury where two- and near three- photon resonances could be combined in order to generate sizeable quantities of photons at $2\omega_1 \pm \omega_2$ (– when $|L_3\rangle$ lies lower than the two-photon resonance). We have done preliminary studies in which 30 µJ of 2525 Å light was combined with 100 µJ of 1.48 Å light to generate ~1.5 nJ at 1164.8 Å. By increasing the input power we shall reach >100 nJ.

REFERENCES

1. G. S. Hurst, M. G. Payne, M. H. Nayfeh, J. P. Judish, and E. B. Wagner, Phys.Rev.Lett., 35:82 (1975).
2. G. S. Hurst, S. L. Allman, M. G. Payne, and T. J. Whitaker, Chem.Phys.Lett., 60:150 (1978).
3. G. S. Hurst, M. G. Payne, S. D. Kramer, and J. P. Young, Rev. Mod.Phys., 51:767 (1979).
4. G. S. Hurst, M. G. Payne, S. D. Kramer, and C. H. Chen, Phys. Today, 33, No.9: 24 (1980).
5. M. G. Payne, C. H. Chen, G. S. Hurst, and G. W. Foltz, in: "Advances in Atomic and Molecular Physics", Vol.17, David R. Bates and Benjamin Bederson, eds., Academic Press, Inc., New York (1981).
6. P. Zoller, Phys.Rev., A 19:1151 (1979).
7. P. Zoller and P. Lambropoulos, J.Phys., B 13:69 (1980).
8. R. Mahon, T. J. McIlrath, V. P. Myerscough, and D. Koopman, IEEE J.Qunatum Electron., QE-15:444 (1979).
9. A. H. Kung, Appl.Phys.Lett., 25:653 (1974).
10. A. H. Kung, J. F. Young, and S. E. Harris, Appl.Phys.Lett., 22: 301 (1973).
11. T. J. McKee, B. P. Stoicheff, and S. C. Wallace, Opt.Lett., 3: 207 (1978).
12. D. Cotter, Opt.Commun., 31:397 (1979).
13. R. Wallenstein and H. Zacharias, Opt.Commun., 32:429 (1980).
14. F. S. Tomkins and R. Mahon, Opt.Lett., 6:179 (1981).

ONE-ATOM DETECTION AND STATISTICAL STUDIES

WITH RESONANCE IONIZATION SPECTROSCOPY*

M. G. Payne and G. S. Hurst

Oak Ridge National Laboratory
Oak Ridge
Tennessee, USA

INTRODUCTION

To learn how to take matter apart atom-by-atom and to count each atom according to its type, regardless of its initial chemical or physical state, is presumably a worthy goal in scientific research[1]. The advent of the laser created real hope that these aspirations will be realized. The counting of atoms is not merely an intellectual exercise set apart from real-world applications. On the contrary, even though the capability is scarcely more than five years old, practical applications have been made in many fields of chemistry, physics, the environment and industry. In this lecture we wish to review how the laser made possible the counting of atoms and how this capability has been put to use in situations where atoms are free to react chemically as they diffuse through a medium. Fluctuation phenomena and statistical mechanics can also be examined in these situations.

ONE-ATOM DETECTION

Both the high selectivity and the extraordinary sensitivity of resonance ionization spectroscopy (RIS)[1] were demonstrated by pulsing a laser directly through a proportional counter (Figure 1). It was shown by Curran et al.,[2] in 1949 that the improved version of the 1908 Rutherford-Geiger electrical counter (now known as a

---

*Research sponsored by the Office of Health and Environmental Research, U.S. Department of Energy under contract W-7405-eng-26 with the Union Carbide Corporation.

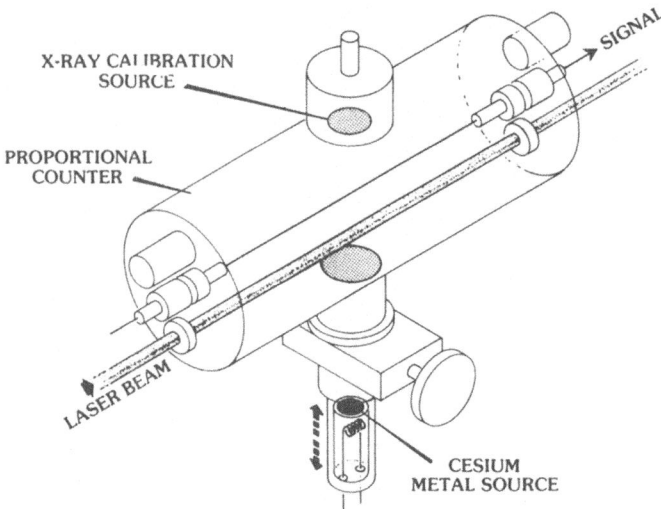

X-RAY CALIBRATION
SOURCE

SIGNAL

PROPORTIONAL
COUNTER

LASER BEAM

CESIUM
METAL SOURCE

Fig. 1. Schematic of an experiment conducted to prove that RIS can
be used to detect a single atom.

proportional counter) can be used to count single electrons at thermal energy. Therefore, when lasers are used to remove one electron
from all of the atoms of a selected type, one-atom detection is made
possible. Proportional counters are normally filled with gases like
Ar (90%) and CH_4 (10%). But, for example, a pulsed laser tuned to
4555 Å can ionize with unit efficiency each atom of cesium without
producing background ionization of the counting gas. In the original
demonstration[3] of one-atom detection at the Oak Ridge National
Laboratory (ORNL) it was proven that one atom of cesium could be
selected out of 10^{19} atoms of the counting gas (Ar and CH_4).

Another important form of atom counting involves the time-
resolved detection of a single daughter atom in flight following the
decay of a parent atom. Thus, it was shown[4] that an individual
atom of cesium could be counted from the fission decay of an individual atom of the isotope ^{252}Cf. The energy released in the fission
process generated a signal in a charged particle detector that triggered the laser used to accomplish the RIS process $Cs(\omega_1, \omega_1 e^-)Cs^+$.
The experiment proved that daughter atoms can be counted in coincidence with the decay of parent atoms. Such techniques could eventually work for most of the daughter atoms associated with radioactive
decay and could possibly be used to reduce backgrounds greatly in
low-level counting facilities.

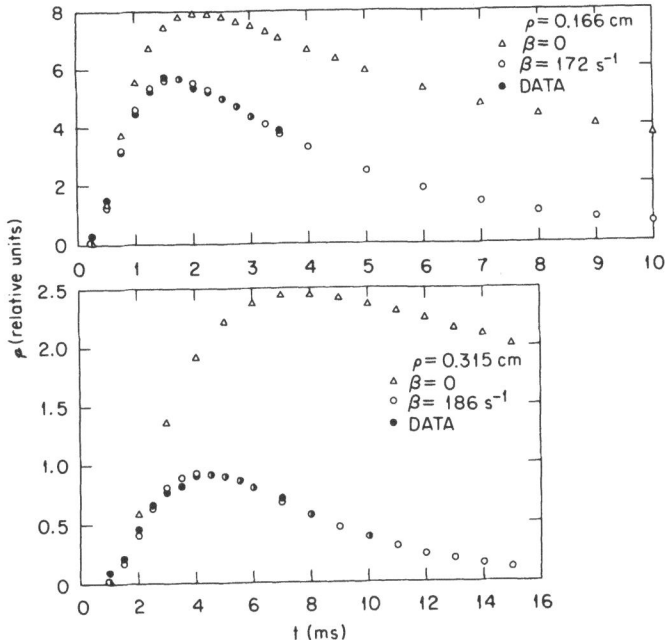

Fig. 2. An example of diffusion data using the RIS detection method
and showing the diffusion of cesium atoms from a line source
to a line detector.

CLASSICAL CHEMICAL PHYSICS APPLICATIONS

The capability for detecting a population of just a few atoms
has made it possible to investigate some problems in classical chemical physics which previously were difficult or impossible. Precision measurements of the diffusion of free atoms amongst other
atoms and molecules have been made in sufficient detail to test the
basic diffusion equation in both time and space domains. The determination of rates of reactions of extremely reactive species (such as
alkali atoms) with other atoms or molecules is now possible. Since
only a few of these reactive atoms need to be produced, several
problems concerning corrosion of the apparatus and the production of
complicated chemical by-products are avoided. Study of a population
of a few atoms (e.g., 10) to observe their statistical behavior has
been made.

These measurements of the diffusion, chemical reaction, and
statistical fluctuation of atoms are best made by starting with a
chemically inert source such as an alkali-halide molecule. Upon
photodissociation of CsI (for example) with a pulsed laser in the
ultraviolet (UV) region, a well-defined source of cesium atoms is
produced along the narrow laser beam at some time, say t = 0. At any

time t > 0 another laser can be pulsed to detect cesium atoms if they exist as free atoms in the path of this detector laser.

For a line geometry the appropriate diffusion equation is

$$\frac{\partial n(\rho,t)}{\partial t} = D\nabla^2 n(\rho,t) - \beta n(\rho,t),$$

where ρ is the distance between source and detector, t is the time, D is the diffusion coefficient, and β is the rate constant for loss due to chemical reactions. The solution for the concentration of free atoms $n(\rho,t)$ is

$$n(\rho,t) = \frac{\lambda}{4\pi Dt} \exp(-\rho^2/4Dt)\exp(-\beta t).$$

for a line source of λ atoms per unit length created at t = 0 and located in an infinite medium. The source and detector can be assumed to be well separated from any surfaces since atoms are both liberated and detected with pulsed laser beams. The data can be analyzed to obtain β and D. We note that the quantity $n(\rho_1,t)/n(\rho_2,t)$ is independent of β but contains D; thus, precision measurements of D can be made, even when the chemically reactive medium has a large β. Experiments are done at two different separations between the source and detector lasers so that chemical effects can be cancelled in the ratio $n(\rho_1,t)/n(\rho_2,t)$. These methods have been used to obtain the diffusion coefficient for cesium atoms and for lithium atoms in several gases or gas mixtures. Besides providing more accurate measurements of the diffusion coefficients for these difficult chemical environments, the method has possible fundamental value. Since both time and space resolution are inherent in the method, it provides a very detailed test of the validity of the diffusion equation itself[5].

Studies have also been made to obtain the reaction constant β in various chemical environments. Thus, Grossman et al.,[6,7] studied the reaction of cesium with O_2 in various atmospheres of both helium and argon. Similar experiments have been reported by Kramer et al.,[8] for lithium atom reactions with O_2 in helium and argon. These were slightly more complex laser experiments since one laser at 2950 Å was used to detect lithium in the RIS scheme $(Li(\omega_1,\omega_2,e^-)Li^+$ with ω_1 corresponding to 6708 Å and ω_2 corresponding to 6104 Å. The source laser and the detector lasers all overlapped in space, this being the geometry in which chemical effects are most easily separated from diffusion losses. Complex pressure-dependent data suggest that the reaction of lithium with O_2 to make stable LiO_2 proceeds both by energy transfer from intermediates like $(LiO_2)^*$ and O_2 reactions with intermediates like LiAr.

A particularly interesting experiment in statistical mechanics has just been completed in our laboratory[9]. Einstein[10] discussed a gedanken experiment to test the ergodic hypothesis for the case of freely diffusing atoms. Essentially, the experimental geometry is the one considered by Einstein except he visualized an infinite plane source and an infinite plane detector instead of a line source and a line detector. In any case, two tests of the solution to the diffusion equation could be visualized. In one case a large number of atoms would be released at t = 0 and the detector would essentially measure the density function n(ρ,t) on each trial for a particular ρ and t. This method was called a space summation method. Einstein's time summation method was visualized as a process where a small number of atoms would be released at t = 0, such that the detector at ρ and t would surely count the rare gas atom arriving at the detector. A series of trials are required so that probabilities are involved in estimating n(ρ,t). If the space summation method and the time summation method agree, the system is ergodic. Our results of an actual experiment are shown in Figure 3. In basic terms, the gedanken experiment visualized by Einstein could actually be done in the laboratory because we can now count individual atoms. Furthermore, the experiment was actually convenient because of a special property of the proportional counter - namely, it is a good digital device for counting one electron (thus one atom, assuming saturated RIS) and a good analog device for summing many electrons (thus many atoms) created by one laser pulse.

ANALYTICAL CHEMISTRY APPLICATIONS

The use of RIS in analytical chemistry has been reviewed by Young et al.,[11] and by Hurst[12]. The essential advantages of RIS to any analytical system are selectivity, sensitivity, and generality to nearly all of the elements.

An example of an analytical application of RIS that requires only lasers for selective ionization and proportional counters for sensitive detection is the development by Mayo et al.,[13] at the National Bureau of Standards (NBS). They have developed Laser Ablation and Resonance Ionization Spectroscopy (LARIS) for the analysis of a few impurity atoms in a solid sample. With LARIS a layer is used to ablate atoms from a solid, followed with a laser for RIS of the neutral atoms. Using a proportional counter as an electron detector, the NBS group was able to detect down to 5×10^{11} atoms/cm^3 of electronics grade silicon, or one atom of sodium in 10^{11} atoms of silicon. Since one electronic device in modern VLSI technology has only 5×10^{10} atoms of silicon, the NBS sensitivity for sodium was such that less than one atom of sodium per device could be detected.

The combination of a RIS ionization source and a mass spectrometer provides the best features of two methods. Essentially, a RIS

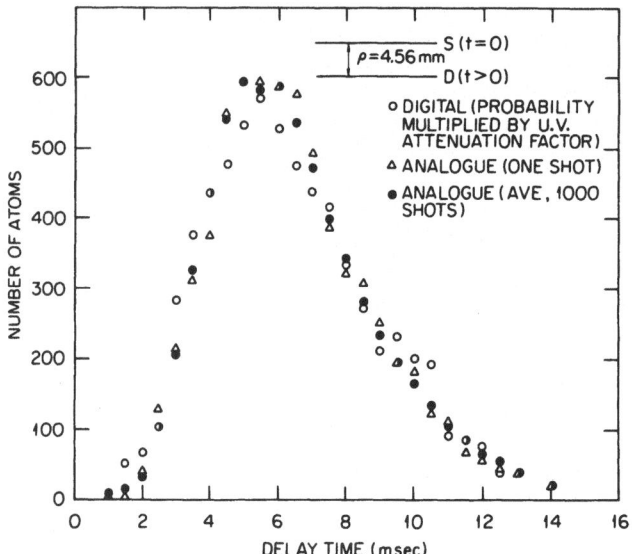

Fig. 3. Experimental data obtained on Einstein's "gedanken" experiment to illustrate the ergodic hypothesis.

source for a mass spectrometer provides selectivity on Z and the mass
analyzer provides selectivity on A. With both Z and A selectivity,
many interferences associated with electron ionization can be eliminated and isobars can be distinguished. The use of the laser as an
ionization source for a mass spectrometer is a very natural idea and
received attention in a 1976 paper by Letokhov[14], who suggested the
combination of lasers and mass spectrometers for more definitive
molecular spectroscopy.

     A particular interesting example of the combination of RIS with
mass spectrometers has been proposed[15,16] for isotopically selective counting of noble gas atoms. Essentially, Maxwell's sorting
demon can be made to work with the intelligence to recognize even an
isotope of an atom and with the memory capacity to store and count
each atom (see next chapter by Payne and Hurst).

REFERENCES

1. G. S. Hurst, M. G. Payne, S. D. Kramer, and J. P. Young, A
     summary and review of the early history of RIS, Rev.Mod.
     Phys., 51:767 (1979).
2. S. C. Curran, A. L. Cockroft, and J. Angus, Phil.Mag., 40:929
     (1949).
3. G. S. Hurst, M. H. Nayfeh, and J. P. Young, Appl.Phys.Lett., 30:
     229 (1977).

4. S. D. Kramer, C. E. Bemis, Jr., J. P. Young, and G. S. Hurst,
    Opt.Lett., 3:16 (1978).
5. G. S. Hurst, S. L. Allman, M. G. Payne, and T. Whitaker, Chem.
    Phys.Lett., 60:150 (1978).
6. L. W. Grossman, G. S. Hurst, M. G. Payne, and S. L. Allman,
    Chem.Phys.Lett., 50:70 (1977).
7. L. W. Grossman, G. S. Hurst, S. D. Kramer, M. G. Payne, and
    J. P. Young, Chem.Phys.Lett., 50:207 (1977).
8. S. D. Kramer, B. E. Lehmann, G. S. Hurst, M. G. Payne, and
    J. P. Young, J.Chem.Phys., 76:3614 (1982).
9. J. Iturbe, S. L. Allman, G. S. Hurst, and M. G. Payne, (to be
    published).
10. A. Einstein, in: "Investigations on the Theory of the Brownian
    Movement", R. Furth, ed., Dover, New York (1956).
11. J. P. Young, G. S. Hurst, S. D. Kramer, and M. G. Payne, Analyt.
    Chem., 51:1050A (1979).
12. G. S. Hurst, Analyt.Chem., 53:1448A (1981).
13. S. Mayo, T. B. Lucatorto, and G. G. Luther, Analyt.Chem., 54:553
    (1982).
14. V. S. Letokhov, in: "Tunable Lasers and Applications",
    A. Mooradian, T. Jaeger, and P. Stokseth, eds., Springer-
    Verlag, Berlin, p.122 (1976).
15. G. S. Hurst, M. G. Payne, S. D. Kramer, and C. H. Chen, Phys.
    Today, 33(9):24 (1980).
16. G. S. Hurst, M. G. Payne, C. H. Chen, R. D. Willis, B. E.
    Lehmann, and S. D. Kramer, in: "Laser Spectroscopy V",
    A. R. W. McKellar, T. Oka, and B. P. Stoicheff, eds.,
    Springer-Verlag, New York/Berlin, pp.59-66 (1981).

WEAK INTERACTION STUDIES USING RESONANCE IONIZATION SPECTROSCOPY*

M. G. Payne and G. S. Hurst

Oak Ridge National Laboratory
Oak Ridge
Tennessee, U.S.A.

COUNTING ATOMS WITH RESONANCE IONIZATION SPECTROSCOPY (RIS)

Important developments in laser sources for the vacuum ultraviolet (VUV) region of the spectrum are making it possible to carry
out resonance ionization [for a review, see Ref.1] of some of the
noble gases. It has already been shown that xenon can be ionized in
a two-photon allowed excitation from the ground state. Recently a
new method of generating radiation by four-wave mixing in mercury
vapor[2] enables excitation of xenon in a one-photon resonance process. With these new laser sources[3] we expect to have effective
ionization volumes of 10^{-3}-10^{-2} cm^3 for the cases of argon, krypton,
and xenon. This has important consequences in weak interaction
physics and environmental research.

Widespread applications of noble gas detectors are due to the
fact that small numbers of the chemically inert atoms can be recovered from very large targets of materials where they may be generated by rare events. In this lecture we show how lasers can be
combined with mass spectrometers to detect a few noble gas atoms of
one isotope in the presence of very large numbers of atoms of a
neighboring isotope. This technique (which we have called Maxwell's
demon because of the atom-sorting functions performed in the apparatus) is described and then followed with a brief discussion of two
applications in weak interaction physics - double-beta decay and the
solar neutrino problem. [Other applications of atom counting have
been reviewed in Ref.4].

*Research sponsored by the Office of Health and Environmental
Research, U.S. Department of Energy under contract W-7405-eng-26
with the Union Carbide Corporation.

DEVELOPMENT OF MAXWELL'S DEMON

Since Resonance Ionization Spectroscopy (RIS) schemes all involve one or more bound-bound transitions, atomic number (Z) selectivity is virtually assured. A RIS source for a mass spectrometer
provides a combination of Z and A (atomic mass) selectivity for the
analysis of individual atoms. This two-fold selectivity can be put
to use in counting atoms of a particular isotope (see Figure 1).

Since the effective volume of the high-vacuum chamber is rather
large (~2.5 liters) and the volume that can be saturated in the
resonance ionization process is quite small (3×10^{-3} cm^3), it is
necessary in a practical application of the one-atom detector to find
a way to increase the chance of a noble gas atom being in the beam at
the time the laser is fired. In our case the sample is frozen onto a
liquid helium cold finger. When the light of a pulsed visible laser
strikes the cold finger, the noble gas atoms are released instantly
and travel a few millimeters before the time-delayed ultraviolet
laser is pulsed just above the cold finger and ionizes a significant
fraction of the sample. The ionization probability for a single atom
of a selected isotope can be adjusted to approximately 10% in order
to achieve true digital counting.

The ability to sort xenon atoms from other atoms in a small
volume by means of selective laser ionization and ion implantation[5]
has been demonstrated[6]. We have shown the depletion of xenon atoms
in volumes of 50 and 15 cm[3], using pulsed dye lasers pumped by
harmonics of a Nd-YAG laser to produce Xe^+ ions followed by implantation of the ions into a channeltron detector. When an ion strikes a
metal, electrons are emitted in sufficient quantity[7] to insure the
counting of each ion.

Since the abundance of sensitivity of the quadrupole mass spectrometer is only 10^3, many atoms of neighboring isotopes will be
implanted for samples in which the selected isotope has a very low
abundance, e.g., $1:10^{15}$. The system, therefore, is designed to allow
for an indium target to be heated or melted inside the high vacuum in
order to release all of the implanted atoms, which will then be
collected again on the liquid helium cold finger. After a small
number, n, of these cycles, an enrichment of 10^{3n} will be achieved
and the selected atoms will be counted by implantation into a copper-
beryllium target.

STUDIES OF DOUBLE-BETA DECAY

Double-beta ($\beta\beta$) within a nucleus can be either a two-neutrino
(2ν) process or a neutrinoless (0ν) process, depending on the nature

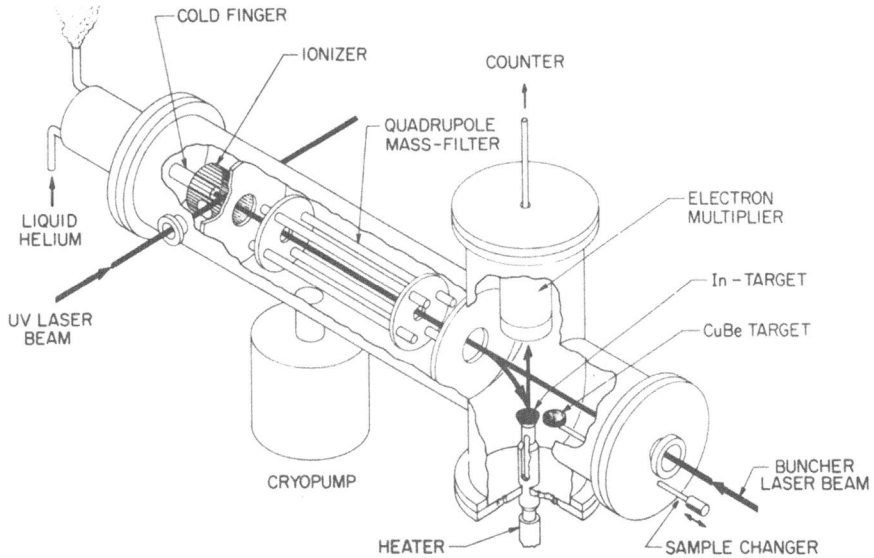

Fig. 1.   Schematic of a method for counting individual atoms of the
          noble gases.   Isotope resolution leads to applications such
          as a ^{81}Kr detector for solar neutrinos on a ^{82}Kr detector
          for $\beta\beta$ decay.

of the neutrino[8].   This is illustrated by starting with the
Majorana neutrino in the Racah sequence:

$$n \rightarrow P + e^- + \nu_M,$$

$$\nu_M + n \rightarrow P + e^-.$$

Such a sequence in a nucleus converts two neutrons to two protons
plus two electrons and is a 0ν process.   If lepton number conservation holds, this sequence does not occur.   Such a (Majorana) particle is identical with its own antiparticle and violates lepton
conservation.   A Dirac neutrino has a distinct antiparticle; hence,
$\beta\beta$ decay must proceed as a 2ν process, i.e.,

$$2n \rightarrow 2p + 2e^- + 2\nu_e$$

if lepton number is conserved.   Rosen argued that the observation of
a 0ν mode in $\beta\beta$ decay would prove that lepton number is not conserved
and would be an extremely important discovery for building grand
unified models.   These are strong motivations for examining the role
that one-atom detection could play in the study of $\beta\beta$ decay.

Kirsten's group[9,10] at Heidelberg made extensive studies of ^{82}Se by detecting ^{82}Kr in geologically old selenium ores. A later result on this total decay rate made by Srinivassan et al.,[11] gave a half-life of $(2.76 \pm 0.88) \times 10^{20}$ yr. On the other hand, in a cloud-chamber experiment, Moe and Lowenthal[12] found a much higher decay rate with $T_{\frac{1}{2}} = (1.0 \pm 0.04) \times 10^{19}$ yr.

The present work could provide a third type of experiment on $\beta\beta$ decay and would fall in the category of a radiochemical tank experiment, similar to the approach which Ray Davis uses for solar neutrino flux measurements. The use of one-atom detection for $\beta\beta$ decay of ^{82}Se could proceed as follows. If 10 kg of natural selenium (9.2% ^{82}Se) is put into a well-evacuated enclosure, 100 atoms of ^{82}Kr will be produced each day if the half-life is 1.5×10^{20} yr. At least once each week an easily measurable signal of ^{82}Kr would be generated. Krypton could be recovered from the selenium and then introduced into the Maxwell demon apparatus for counting each atom.

Recent theoretical results for the $\beta\beta$ decay in ^{128}Te and ^{130}Te also show that the rates are much greater than the geochemical values. Thus, new experiments on tellurium isotopes are especially attractive as a check on the geochemical measurements. Furthermore, Rosen[8] points out that if nuclear matrix elements are assumed to be equal for the transitions ^{128}Te $\rightarrow$ ^{128}Xe and ^{130}Te $\rightarrow$ ^{130}Xe, then the ratio of their lifetimes just depends on phase space considerations. This argument is used to predict that the ^{128}Te lifetime is 200 times longer than the ^{130}Te lifetime if the decay is pure $2\nu\beta\beta$; whereas if pure $0\nu\beta\beta$, the same lifetime ratio is 6×10^3.

The new kind of $\beta\beta$-decay experiment based on counting the atoms should apply to the tellurium cases. Assume, for example, that $T_{\frac{1}{2}} = 3 \times 10^{24}$ yr for ^{128}Te and 3×10^{21} yr for ^{130}Te. A 200-liter tank filled with 1000 kg of tellurium would produce one atom of ^{128}Xe per day and 1000 atoms of ^{130}Xe per day. Such an experiment could reveal whether the neutrino is a Dirac or a Majorana particle.

SOLAR NEUTRINO FLUX DETERMINATION

Neutrinos are the only particles that are able to escape from the center of the sun where thermonuclear processes occur. They thus serve as the only direct means of testing the standard theory of solar fusion processes[13]. Recent measurements[14] with a ^{37}Cl$(\nu,e^-)^{37}$Ar detector indicate that the neutrino flux due to the ^{8}B source in the sun is less than one-third of that expected from standard models. This discrepancy can apparently only be resolved by rather drastic changes in the solar model or in weak-interaction theory[15]. Due to the large discrepancy between the only available measurements and theory, it is highly desirable to perform additional experiments.

Fig. 2.   Photograph of the large tank used by Ray Davis Jr., to study
the flux of solar neutrinos on the earth[15].

The present development of a counter of noble gas atoms with
isotopic selectivity should make it possible to perform a significant
variation of the Ray Davis chlorine experiment. Imagine that the
large tank shown in Figure 2 is filled with ethylene dibromide
(CH_2BrCH_2Br) instead of C_2Cl_4. The neutrino capture $^{81}Br(\nu,e^-)^{81}Kr$
(due to ^7Be solar neutrinos) can then be used as a detector provided
that the 2.1×10^5-yr ^{81}Kr can be counted. Clearly, since only 500
atoms of ^{81}Kr are produced in 862 tons of ethylene dibromide in one
year, decay counting is not possible. The success of such an experiment will depend on our ability to count about 100 atoms of ^{81}Kr in
the presence of 10^8 atoms of ^{82}Kr which could be collected from the
tank due to atmospheric contamination. New experiments on solar
neutrinos, neutrino oscillation, and $\beta\beta$ decay are being implemented
to test fundamental assumptions in weak interaction physics. These
depend on ultrasensitive analytical methods made possible by lasers.

REFERENCES

1. G. S. Hurst, M. G. Payne, S. D. Kramer, and J. P. Young,
      Rev.Mod.Phys., 51:767 (October 1979).

2. F. S. Tomkins and R. Mahon, Opt.Lett., 6(4):179 (1981).
3. Yun-Mui Yiu, K. D. Bonin, and T. J. McIlrath, Opt.Lett.,
    7(6):268 (1982).
4. G. S. Hurst, M. G. Payne, S. D. Kramer, and C. H. Chen, Phys.
    Today, 33(9):24 (1980).
5. R. G. Wilson and G. R. Brewer, in: "Ion Beams", Wiley, New York,
    (1973).
6. C. H. Chen, G. S. Hurst, and M. G. Payne, Chem.Phys.Lett., 75:
    473 (1980).
7. K. H. Krebs, Fortschr.Phys., 16:419 (1968).
8. S. P. Rosen, in: "Double Beta Decay and Limits on the Mass of
    Majorana Neutrinos", Presented at the Workshop on Neutrino
    Masses, October 2-4, 1980, Cable, Wisconsin.
9. T. Kirsten, W. Gentner, and O. A. Schaeffer, Z.Physik, 202:
    273 (1967).
10. T. Kirsten and H. W. Mueller, Earth Planet.Sci.Lett., 6:271
    (1969).
11. B. Srinivassan, E. C. Alexander Jr., R. D. Beaty, D. E. Sinclair
    and O. K. Manuel, Econ.Geol., 68:252 (1973).
12. M. K. Moe and D. D. Lowenthal, Phys.Rev., C 22:2186 (1980).
13. J. N. Bahcall and R. L. Sears, Ann.Rev.Astron.Astrophys., 10:25
    (1972).
14. R. Davis Jr., and J. M. Evans, in: "Proceedings 13th Cosmic
    Radiation Conference", Vol.3, p.2001 (1973).
15. J. N. Bahcall, Rev.Mod.Phys., 50:881 (1978).

RESONANCE IONIZATION SPECTROSCOPY:

APPLICATIONS IN ISOTOPE GEOPHYSICS

B. E. Lehmann

Physics Institute University of Bern
Switzerland and Oak Ridge National Laboratory
Oak Ridge, Tennessee, USA

INTRODUCTION

A quantitative understanding of processes in our global environment is to a large extent based on the analysis of isotope ratios. Isotopes of interest very often occur in natural samples such as air, water, ice, sediments or organic material only in extremely small concentrations and very sensitive analytical techniques are necessary to extract the information.

Resonance Ionization Spectroscopy (RIS) combines the selectivity of optical spectroscopy with the sensitivity of ionization detectors to a powerful new analytical method and enables the counting of single atoms of a selected element in backgrounds of other atoms or molecules as high as 1 in 10^{19}[1,2].

The goal of this lecture is

a) to outline the methods and problems of isotope geophysics,
b) to discuss the capabilities of RIS as a new analytical technique for this field,
c) to present research which is in progress at the Oak Ridge National Laboratory (USA) and at the University of Bern (Switzerland) where the RIS technique is combined with a Quadrupole Mass Spectrometer to count single atoms of radioactive noble gases such as ^{81}Kr or ^{39}Ar for dating applications.

ISOTOPE GEOPHYSICS

Every sequence of samples from an environmental system contains information about processes that occur in nature. Isotope concentrations in these samples are fingerprints of the dynamics of atmosphere, hydrosphere and biosphere. Isotopes of potential interest include stable as well as radioactive isotopes with quite different halflives. Accordingly information on time scales from a few days (d) up to several hundred thousand years (a) is attainable.

The following is a list of selected isotopes and examples of current research in this field. The list of course is by no means complete[3].

^3H (12.3a)

Rainwater concentrations reflect seasonal variations in the mixing of bomb-produced ^3H from the stratospheric reservoir into the troposphere.

^{10}Be (1.6×10^6a)

The history of solar activity can be reconstructed by ^{10}Be measurements on polar ice cores since ^{10}Be production by cosmic rays in the earth atmosphere is modulated by solar magnetic fields.

^{13}C (stable, 1.1%)

A reconstruction of $^{12}C/^{13}C$-ratios in atmospheric Co_2 gives information about past biospheric activity because CO_2 released from plants differs in ^{13}C (by a very small fraction) from CO_2 in the atmosphere due to fractionation in the assimulation process.

^{14}C (5730a)

The best known and most widely used isotope for dating samples for archaeology, palaeobiology and related sciences but also a major isotope to understand the global CO_2-cycle with its impact on man (greenhouse effect).

^{18}O (stable, 0.204%

Deviations of the $^{16}O/^{18}O$-concentration from a standard value in e.g. Greenland deep drilling ice cores yield information about past climatic conditions as far back as into the last ice age.

^{36}Cl (3×10^5a)

Cosmic-ray produced ^{36}Cl continuously gets attached to aerosols, is washed out and enters the hydrosphere. It may become a very valuable tracer in groundwater hydrology.

^{37}Ar (34.8d)

Atmospheric nuclear weapons tests represent an input of ^{37}Ar well defined in space and time. ^{37}Ar in air samples from different locations on earth taken in the days and weeks after the tests can be used to learn about global atmospheric circulation.

^{39}Ar (269a)

The ideal tracer for deep sea circulation studies. Its halflife fits very well to the time range of mixing processes in the oceans.

^{81}Kr (2.1×10^5a)

A cosmic-ray produced noble gas radioisotope that is a perfect candidate for dating very old groundwater on a time scale that is important in connection with the selection of sites for nuclear waste deposition in geologic formations.

^{85}Kr (10.72a)

Plutonium production for military purposes and the commercial use of nuclear energy are responsible for a more or less linear increase of ^{85}Kr in the atmosphere over the past 20 years. In iso-tope geophysics air-sea exchange processes may e.g. be studied based on ^{85}Kr measurements.

^{133}Xe (5.25d)

Microclimatic models can be developed and tested using ^{133}Xe released from nuclear power plants as a local tracer.

^{210}Pb (22.3a)

In the natural decay series of uranium the short-living ^{222}Rn escapes as a gas from the earth crust; its solid decay-product ^{210}Pb precipitates back to the surface. The study of glacier dynamics and of sedimentation rates in lakes, rivers and estuaries are among the applications of this isotope in geophysics.

The analytical techniques that are used to measure these iso-topes in natural samples are conventional mass spectrometry for stable isotope analysis (^{12}C/^{13}C or ^{16}O/^{18}O) and various types of decay counting systems such as gas proportional counters, liquid scintillation counters or semiconductor counters for the radioactive isotopes. Furthermore advanced accelerator-based mass spectrometry has been introduced to isotope geophysics for isotopes such as ^{14}C, ^{10}Be or ^{36}Cl.

Stable isotope measurements have to be very accurate; deviations of the isotope ratio of 0.1% or less from a standard value are measured on a routine basis.

Radioisotope measurements on the other hand require extremely sensitive analytical methods. Most advanced low level counting facilities have been built in recent years.

To illustrate the state of the art in this field we consider a ^{39}Ar-dating of a deep ocean water sample.

The natural ^{39}Ar-activity of a modern air sample is only 0.1 dpm (decays per minute) per liter of argon. This corresponds to an ^{39}Ar/Ar-ratio of 8 x 10^{-16}. Samples of 300 cc STP Ar can be measured in small high pressure gas proportional counters of 16 cm^3 volume that have a background count rate of only 0.03 cpm (counts per minute). Such extremely low values can be achieved in an underground laboratory built of special low activity concrete with counters constructed out of high purity copper that are operated inside a massive lead shielding and with anticoincidence arrangements[4].

In spite of all this effort the counting time for one sample is still 6 weeks and 1000 liters of ocean water is the minimum sample size.

A fundamental step forward in the analytical methods would be the direct counting of the atoms of the selected radioisotope without waiting for its decays. This of course is especially true for the radioisotopes with very long halflives.

As mentioned before this is now possible for certain isotopes with the new accelerator based mass spectrometry where ^{14}C, ^{10}Be or ^{36}Cl atoms can be counted in samples of polar ice for example in concentrations of 10^5-10^7 atoms per one kilogram of ice.

Among all the radioisotopes of interest in isotope geophysics the noble gases have the distinct advantage of being chemically inert. Unfortunately a tandem-accelerator operates with negative ions and therefore this new instrument is not suitable for noble gas analysis.

We will now discuss how resonance ionization spectroscopy can be used in isotope geophysics with radioactive noble gases. Figure 1 is a further illustration of ongoing research in this field.

RIS FOR NOBLE GASES

Resonance Ionization Spectroscopy (RIS) combines the excitation and ionization of selected atoms by pulsed tunable lasers with highly

sensitive detectors for ions or electrons such as proportional
counters, channeltrons or electron-multipliers. The principle and
some of the first experiments as well as a number of applications in
physics and chemistry are presented in other lectures of this course.

The RIS-technique is basically very general and can be used to
detect atoms of any element in the periodic table (except He and Ne)
by tuning lasers to the corresponding wavelength for selective excitation. In praxis however there are large differences between the
schemes for different elements. The situation is illustrated in
Figure 2. For the elements on the left side of the periodic table
with low ionization potentials such as the alkali group rather simple
schemes can be found. Cesium for instance can be excited and ionized
with one single dye laser tuned to the 4593Å transition in Cs.
Volumes of several cm^3 may be completely saturated in one single
pulse. Noble gases have very much higher ionization potentials and
the lowest levels for optical excitation are considerably higher as
compared to Cs. One photon excitation for Xe, Kr and Ar is only
possible with light in the VUV-region (below 2000Å). Even for a
two-photon excitation ultraviolet laser pulses are necessary.

RIS for Xenon was demonstrated with a Nd:YAG pumped dye laser
system that produced 1 mJ pulses of about 4 nsec duration at the
2526Å and 2496Å two-photon resonance transitions of Xe[5]. The light
was focused to a power density of 10^9 W/cm^2. Due to the low cross
section for two-photon-absorption the effective volume of ionization
is only on the order of 10^{-4} cm^3 per pulse.

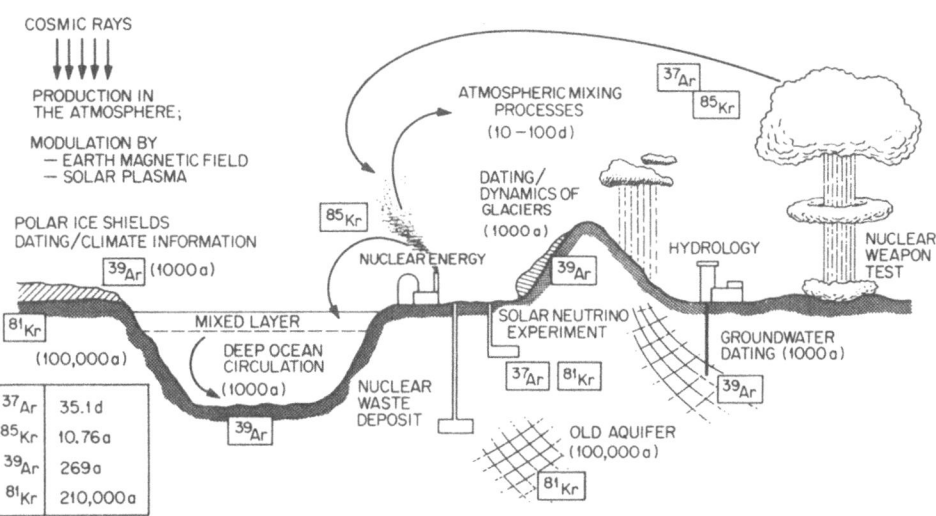

Fig. 1. Radioisotopes of Argon and Krypton in Isotope Geophysics.

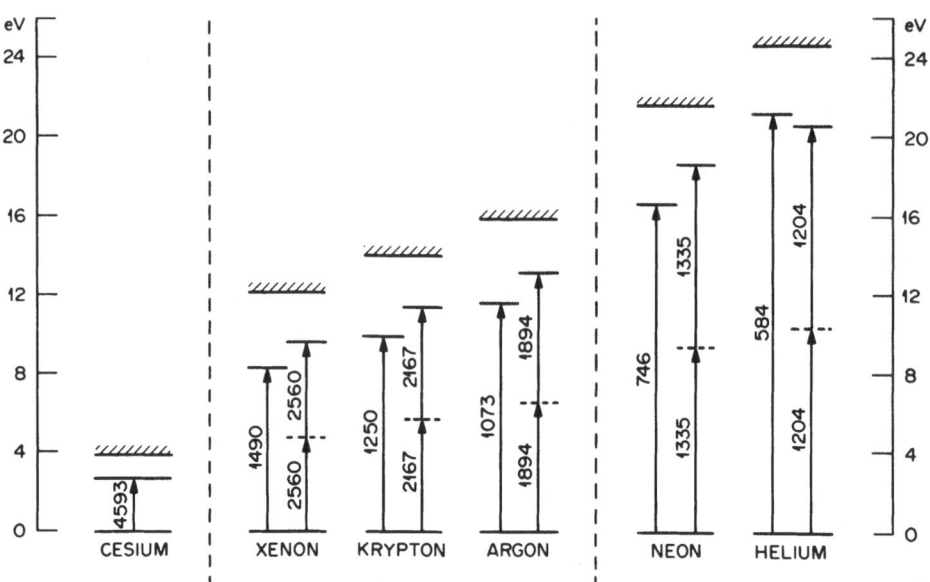

Fig. 2.   RIS excitation schemes for noble gases as compared to Cs.

The situation for Kr and Ar is even worse.   RIS for Kr was demonstrated with a XeCl excimer laser pumped dye laser system that produced light at the 2167Å two-photon resonance transition of Kr[6].

No RIS-work on Ar has been reported so far.

Two developments in laser technology have influenced RIS for heavy noble gases in the last one or two years:

1) high-power, narrow-band excimer laser UV sources have become available.
2) tunable coherent light sources in the VUV have been constructed by using 4 wave-mixing techniques.

The situation is illustrated for Krypton in Figure 3.

The 6 p level in Krypton can - by coincidence - be excited by two photons from a ArF* excimer laser at 1927.5Å. Commercial systems generate pulses with high power (> 100 mJ) and high repetition rate (100 pps) but with large bandwidth (100 cm^{-1}). By using advanced injection locking techniques most of the energy can be concentrated into a bandwidth as small as 0.01 cm^{-1}[7]. Such a tunable high spectral brightness source is an ideal instrument for the selective ionization of Kr. Unfortunately the system is by no means simple or inexpensive.

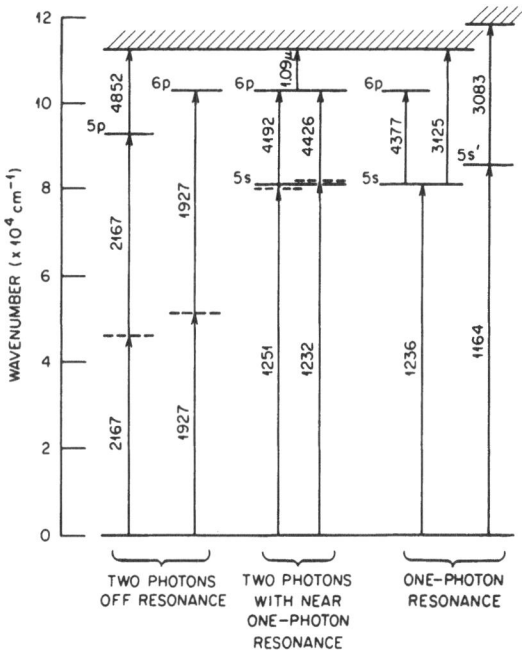

Fig. 3. RIS schemes for Krypton.

Several other RIS schemes for Ar, Kr and Xe have been proposed based on existing data of commercially available excimer laser systems[8].

Another approach became possible through progress in VUV-generation by 4-wave mixing[9]. Coherent tunable radiation around 1251Å and 1232Å can be produced in Hg-vapour.

These wavelengths are very close to the lowest one-photon allowed transition in Kr and RIS-schemes with two-photon excitation enhanced by this near one-photon resonance are possible (see Figure 3).

Similar 4-wave mixing schemes in Xenon are under investigation with the final goal of generating tunable light for one-photon excit-ation of Kr at 1236Å and Ar at 1066Å.

Isotope effects in noble gas spectra are very small. Isotopic selectivity in the range of 10^{-12} to 10^{-16} is required in appli-cations for isotope geophysics. This cannot be achieved by optical isotope selective excitation. Therefore element-selective laser ionization is currently combined with a quadrupole mass spectrometer to an instrument that will allow the counting of atoms of any sel-ected isotope of the heavy noble gases.

This instrument is described in the next section.

ISOTOPE SELECTIVE NOBLE GAS ATOM COUNTING

Figure 4 is a schematic drawing of a system that was built at the Oak Ridge National Laboratory[10]. It is planned to use this machine for ^{81}Kr dating of very old water samples.

We will discuss the analytical procedure step by step:

1) One liter of water contains about 6×10^{-5} cc STP Kr. The Kr/^{81}Kr ratio in a modern air sample is 1.67×10^{12}. A Kr sample extracted from one liter of water therefore contains about 1000 ^{81}Kr atoms in a background of more than 10^{15} atoms of the stable Kr atoms. Such a sample is introduced to the quadrupole system of 4000 cm^3. It will generate a total pressure of about 10^{-5} Torr.

2) Since even the best available RIS schemes for Kr can only ionize small volumes per pulse it is essential to concentrate the sample in the RIS beam at the time when the laser fires. An "atom buncher" had to be invented. It combines cryogenic cooling and trapping of noble gas atoms to a well defined liquid helium cooled cold spot with pulsed laser surface heating. The Kr sample is now trapped on the surface of the cold finger.

3) A 1-μsec pulse from a commercial dye laser fired down the axis of the quadrupole mass spectrometer will heat this surface within a very short time and the Kr atoms will leave the surface immediately.

4) The RIS lasers are fired with a time delay of about 20 μsec and will ionize a few percent of the sample in one pulse. All isotopes are ionized simultaneously and enter the quadrupole system which is tuned to mass 81.

5) The ^{81}Kr atoms pass the quadrupole and are implanted by a high electric field into a target. Isotopes with a different mass (e.g. ^{80}Kr and ^{82}Kr) have only a probability of 10^4:1 to pass the quadrupole system. Therefore, after a certain number of pulses essentially all ^{81}Kr isotopes are implanted into the target. The number of other isotopes of Kr is reduced by a factor of 10^4.

6) The rest of the gas is now pumped off. The target contains 1000 ^{81}Kr and about 10^{10} ^{82}Kr atoms. It can be heated to release the implanted atoms. They will undergo a second and third enrichment cycle.

7) The final counting of ^{81}Kr will be done either by counting secondary electrons emitted from the target or by direct ion counting with the electron multiplier.

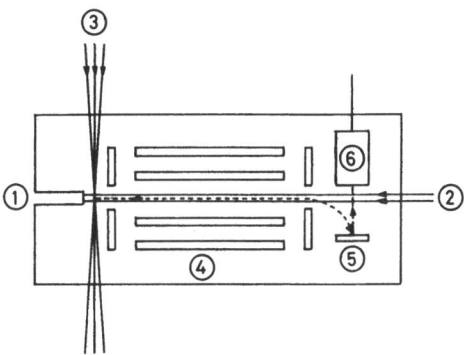

Fig. 4.   RIS in a masspectrometer. Schematic arrangement with 1)
          Cold Finger 2) Buncher Laser 3) RIS laser 4) Quadrupole Mass
          Filter 5) Target for Implantation 6) Electron
          Multiplier.

SUMMARY

     The combination of Resonance Ionization Spectroscopy with Mass
Spectrometry can become an extremely valuable new analytical tool in
isotope geophysics. It will help to gain new information about human
impacts on our global environmental system. ^{39}Ar-studies to better
understand deep ocean circulation and the ocean's role as a sink of
man-made CO_2 and ^{81}Kr-dating of old groundwater in connection with
the selection of nuclear waste disposal sites in geological formations are two examples of current interest in this field.

Acknowledgements

     Members of the group at the Oak Ridge National Laboratory that
contribute to the RIS noble gas project are G. S. Hurst, M. G. Payne,
C. H. Chen, R. D. Willis, S. D. Kramer, S. Allman and R. Phillips.
At the University of Bern H. Oeschger and H. H. Loosli are involved
in this project.

REFERENCES

1. G. S. Hurst, M. H. Nayfeh, and J. P. Young; A demonstration of
    one-atom detection, Appl.Phys.Letters, Vol. 30, 5:229
    (1977).
2. G. S. Hurst, M. G. Payne, S. D. Kramer, and J. P. Young,
    Resonace Ionization Spectroscopy and one-atom detection.
    Rev.Mod.Phys., Vol. 51, 4:767 (1979).
3. H. Oeschger, The contribution of radioactive and chemical dating
    to the understanding of the environmental system, in:

"Proceedings of the Symposium on Nuclear and Chemical Dating Techniques, 179th Meeting of the American Chemical Society, Houston", ACS Symposium Series 176, (March 1980).

4. H. H. Loosli, M. Heimann, and H. Oeschger, Low level gas proportional counting in an underground laboratory, Radio-carbon, Vol. 22, 2:461 (1980).

5. C. H. Chen, G. S. Hurst, and M. G. Payne, Direct Counting of Xe atoms, Chem.Phys.Lett., Vol. 75, 3:473 (1980).

6. B. A. Bushaw and T. J. Whitaker, Doppler-cancelled two photon resonant ionization spectroscopy, J.Chem.Phys., Vol. 74, 11:6519 (1981).

7. H. Egger, T. Srinivasan, K. Hohla, H. Scheingraber, C. R. Vidal, H. Pummer, and C. K. Rhodes, A tunable ultrahigh spectral brightness ArF* excimer laser source. Appl.Phys.Lett., Vol. 39, 1:37 (1981).

8. M. G. Payne, C. H. Chen, G. S. Hurst, S. D. Kramer, W. R. Garrett, and M. Pindzola, RIS schemes for Ar, Kr and Xe, Chem.Phys.Lett., Vol. 79, 1:142 (1981).

9. F. S. Tomkins and R. Mahon, High efficiency four wave sum and difference mixing in Hg vapour, Optics Lett., Vol. 6, 4:179 (1981).

10. G. S. Hurst, M. G. Payne, C. H. Chen, R. D. Willis, B. E. Lehmann, and S. D. Kramer, RIS: Counting Noble Gas Atoms, in: "Proceedings of the Fifth International Conference on Laser Spectroscopy", Jasper Park, Alberta, Canada, Springer-Verlag, (June 1981).

ANALYTICAL OPTOGALVANIC SPECTROSCOPY IN FLAMES

J. C. Travis

US National Bureau of Standards
Washington, D.C.
USA

INTRODUCTION

Optogalvanic spectroscopy is restricted by its nature to free atoms or molecules in weakly ionized plasmas, such as flames and low pressure glow discharges. In such environments, a low degree of ionization is maintained by collisions between atoms, molecules, ions, and/or electrons. The use of a laser to produce significant perturbations in the steady state energy level population distribution of a neutral species in a weakly ionized plasma will result in a corresponding change in the volume ionization rate of the plasma. Such a change in plasma impedance in response to discrete optical electronic excitation of atomic or molecular species is called the optogalvanic effect[1,2], and forms the basis of optogalvanic spectroscopy[3-5].

When defined as above, modern optogalvanic spectroscopy has its roots in the work of Penning in 1928[6]. Given a somewhat broader definition, optogalvanic spectroscopy embraces a significant body of literature employing thermionic diodes and extending from 1925[7] to the present[8]. However, despite these roots significantly predating lasers, the tunable laser is undoubtedly responsible for the present activity in the field, which has seen over 100 laser-based publications in the period from 1976 to the present[9].

The bulk of the literature covering optogalvanic spectroscopy concerns low pressure gas discharges. However, virtually all of the research in the analytical application of the optogalvanic effect has utilized the atmospheric flame[2,10]. Unlike the flow discharge, the atmospheric pressure flame exhibits local thermodynamic equilibrium. For this reason, laser excitation of species in flames invariably

213

results in _increased_ ionization (or decreased impedance) of the flame, whereas impedance changes of either sign are found to occur in glow discharges. Thus, the descriptive name "laser enhanced ionization" (LEI)[11], which has been used for the optogalvanic effect in flames in most of the analytical studies to date, is appropriate only for the flame subset of the optogalvanic effect.

Since the flame is a chemically sustained plasma (as contrasted to the electrically-sustained flow discharge), a system of electrodes in an appropriate external circuit must be added to detect the LEI signal. Normally, a water-cooled, negatively-biased, high voltage electrode is immersed in the flame about 2 cm above the burner head. Analyte atoms are selectively excited by one or two pulsed tunable lasers in the irradiated volume just below the cathode. Increases in the ionization rate due to laser excitation are observed as pulsed increases (synchronous with the laser) in the current passed through the flame and measured at the burner head (anode). The amplitude of these current-increase pulses yields a quantitative measure of the analyte concentration in a solution aspirated into the flame in the manner common to flame spectrometric methods.

The following discussion treats the theoretical and experimental development of LEI in flames in some detail, as befits its established analytical relevance. However, the potential for the development of analytical methods based on the optogalvanic effect in other media should not be overlooked.

PHYSICAL MECHANISMS IN FLAMES

Ionization Enhancement

Assuming the existence of local thermodynamic equilibrium in an atmospheric pressure flame, the probability (P_j) that an atom – of ionization potential E_i and occupying an energy level j an energy E_j above the ground state – will be ionized in any given collision is given by the Arrhenius expression:

$$P_j = \exp[-(E_i-E_j)/kT] \tag{1}$$

where k is Boltzmann's constant, and T the flame temperature. For typical flame temperatures of 2000°K < T < 2500°K, kT $\approx$ 0.2 eV is much smaller than typical ionization potentials of 5 eV < E_i < 10 eV, so that P_j is exceedingly small for all but the highest lying energy levels. Equation (1) may be used to show that a 1 eV increase in E_j yields an increase of about 2 orders of magnitude in P_j at flame temperatures. Thus the use of a photon of laser light of energy 2 eV < E_λ < 5 eV to promote an atom to a higher energy level will enhance the ionization probability per collision by 10^4 – 10^{10}.

For a more macroscopic description, the volume ionization rate of a given species in a flame (ions generated per cm^3 per sec) is given by the rate equation

$$\frac{dn}{dt}+ = \sum k_j n_j \simeq \alpha n_+ n_-$$ (2)

where n_+, n_-, and n_j are number densities (number per cm^3) of ions of the species, electrons, and free atoms of the species in state j, respectively; the k_j are ionization rate constants (sec^{-1}) for each state, j; and α is the recombination coefficient ($cm^3 sec^{-1}$). In the absence of laser perturbation, the n_j reflect the Boltzmann distribution, with the majority of the population occupying low-lying states ($E_j < kT$). Even so, many terms of (2) can be important, because the ionization rate constants,

$$k_j = \nu_j P_j$$ (3)

favor higher lying states through the probability P_j, see Equation (1). The collision frequency, ν_j (sec^{-1}), for atoms in state j colliding with any species in the flame, may also increase with excitation, since excited atoms increase in size (and, hence, collision cross section)[12].

When a laser is used to significantly increase the population of a state, m, over its normal value, the term $k_m n_m$ becomes the dominant one in the summation in (2), as well as dominating the recombination term:

$$\frac{dn}{dt}+ \simeq k_m n_m.$$ (4)

he laser-maintained excited atom number density may be described by

$$n_m = F(n_T - n_+),$$ (5)

here n_T is the total number density of the species (so that $n_T - n_+$ is the number density of neutrals remaining), and F is the fraction of neutral atoms maintained in state m by the laser. Substituting (5) into (4) and integrating (4) over the laser pulse duration, τ_L, assuming a rectangular temporal profile (F = constant for $0 \le T \le \tau_L$, F = 0 otherwise) gives the number density of ions created by the laser pulse:

$$n_+(\tau_L) = n_T[1 - \exp(-k_m F \tau_L)].$$ (6)

From equation (6) it may be seen that complete ionization may be approached ($n_+ \rightarrow n_T$) if $k_m F \tau_L$ can be made $\gg 1$. Such conditions may indeed be achieved. Modern pulsed lasers are readily capable of

driving many atomic transitions to the point of "optical saturation",
for which

$$F_s = \frac{g_m}{g_0 + g_m} ,$$

(7)

here F_s is equal to F at saturation, and g_0 and g_m are the statistical weights of the ground state and state m. Thus, a typical pulse
length of $\tau_L \approx 10^{-8}$ sec and excitation fraction of $F \approx \frac{1}{2}$ could yield
near-total ionization for $k_m >> (0.5 \times 10^{-8} \text{sec})^{-1}$, or $k_m >> 2 \times 10^8$-
sec^{-1}. For a reasonable collision rate of $\nu_m \approx 10^9 \text{ sec}^{-1}$, equation
(3) shows that a probability $P_m > 0.2$ would be required. Thus, from
equation (1), m would need to be a state for which $(E_i - E_m) < -\ln(0.2) \cdot$
kT, or $E_i - E_m \approx 0.3$ eV for kT ≈ 0.2 ev.

In practice, complete ionization seems to be easier to achieve
than indicated above, probably because collision rates of excited
states may greatly exceed the commonly assumed value of $\nu_0 \approx 10^9 \text{sec}^{-1}$
for ground state species. Nevertheless, the need to populate a level
within 1 eV or so of E_i is often beyond the reach of a single laser,
and stepwise excitation is invoked[13,14].

Most of the discussion to this point applies strictly only to
the resonant (R) excitation scheme shown in Figure 1 for In. Stepwise excitation (S) is utilized for elements of higher ionization
potential, for which a single step excitation is unable to populate a
level close enough to E_i for efficient ionization. The practical
consequence of stepwise excitation for Equation (6) is that F becomes
a more difficult quantity to calculate, requiring a three-level
steady state rate equation analysis. However, to a first approximation, the composite F may be considered a product of F values for
each of the two steps. Thus, if each transition can be driven to
optical saturation, the composite F will be of the same order of
magnitude as for the single step case. However, below optical saturation, the composite F would shrink drastically, making stepwise
excitation impractical for cw LEI.

Under limited circumstances, both excitation schemes discussed
above may be initiated from low-lying excited states, instead of the
ground state. Such "non-resonant" schemes complete the set of four
possible excitation processes for LEI as shown in Figure 1. Complete
ionization with such non-resonant schemes requires the collisional
mixing of the bottom two states to be fast with respect to the laser
pulse length. For 5-nsec laser pulses, states below $\sim 2000 \text{ cm}^{-1}$ seem
to satisfy this criterion.

Although stepwise excitation is problematic for low power cw
lasers, such lasers are capable of achieving near-total ionization
for low ionization potential elements ($E_i < 6$ eV). The typical
population fraction of $F \approx 0.01$ is more that offset in Equation (6) by

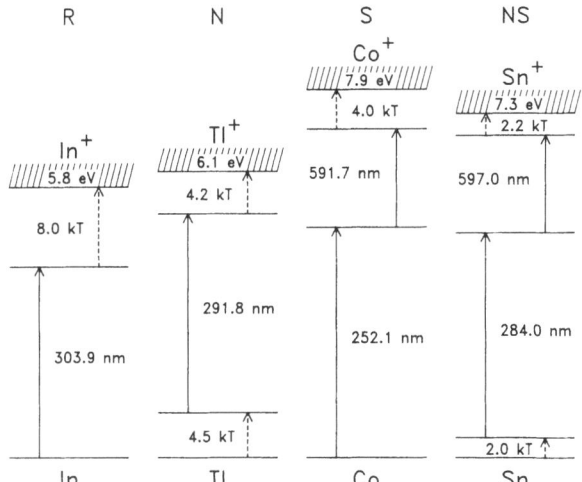

Fig. 1.   Examples of four LEI excitation schemes in a 2500 K flame:
          R = resonant (ground state connected); N = nonresonant; S =
          stepwise; and NS = non-resonant, stepwise.

the increase in τ_L. Indeed, since the laser is continuous, the
interaction time is no longer a pulse length, but the transit time
required for an atom to cross the laser beam travelling at the burnt
gas velocity of the flame (∿10 m/sec)[15].

When ionization is driven nearly to completion, the original
atom density is nearly depleted. This neutral atom depletion can be
observed and quantitated by a straightforward absorption experiment.
Depletions up to 75 percent have been observed for cw excitation of
Na in a C_2H_2/air flame[15]. Discrete, atom-depleted "holes" can be
generated with 1-msec "pulses" derived from the cw laser with an
acousto-optic modulator. The passage of these holes can be observed
in absorption for a few cm above the ionization volume[15]. The
persistence of the holes illustrates the minor role of ion-electron
recombination in the flame, as assumed in the derivation of Equation
(6). It also illustrates the minor role of diffusion, further discussed below.

Ion Transport in Flames

If an LEI ion/electron pair is created in the presence of an
electric field, the ion will experience an electrostatic force in the
direction of the field, and the electron will experience an equal
force in the opposite direction. Since the flame provides
"frictional" resistance to the motion of the charges, each achieves
a constant velocity[16]

$$\vec{v}_+ = -\mu_+\vec{E} \tag{8}$$

$$\vec{v}_- = \mu_-\vec{E} \tag{9}$$

determined by its mobility, μ, where subscripts + and - denote ions and electrons, and E is the electric field. The sign convention used in (8) and (9) is not the conventional one, but is adopted to simplify ensuing equations. Mobilities used here are positive definite quantities, with the direction of motion denoted by the explicit sign. The reason ions move with negative velocities and electrons with positive velocities is that we choose to measure distance from the cathode instead of the anode. Hence, the unit vectors for field and velocity are antiparallel. The mobility is a function of the temperature, pressure, and composition of the flame, and the mass of the transported charged species. Electrons are 100 - 1000 times more mobile than ions[16].

One of the consequences of the great difference in electron and ion mobilities is the modification of the electric field distribution in the flame from what would be expected in air or vacuum. Lawton and Weinberg[16] have developed a simple model to illustrate this behavior for the "one-dimensional" case of a uniform flame between plane parallel electrodes of infinite extent. Several results from this model are shown in Figure 2. The dashed line illustrates the field of 500 V cm^{-1} that would be expected from the application of 1000 V across 2 cm of air.

In the presence of a flame, however, the field magnitude is given by

$$E(x) = \delta r^{1/2}(Y-x), \ x<Y; \ E(x) = 0, \ x>Y \tag{10}$$

$$Y = [2V_a/\delta]^{1/2}r^{-1/4} \tag{11}$$

$$\delta \equiv [e/(\mu_+\epsilon_o)]^{1/2} \tag{12}$$

where r is the volume ionization rate of the flame (ions generated per cm^3 per sec), V_a is the voltage applied across the flame (volts); $\epsilon_o = 8.85 \times 10^{-14}$ F/cm is the permittivity of free space; e is the charge on an electron (coulombs); and x is the spatial coordinate measured from the cathode (cm). The field magnitude may be seen to decrease from a maximum value at the cathode,

$$E(x=0) = [2V_a\delta]^{1/2}r^{1/4}, \tag{13}$$

to zero at x = Y. The region of non-zero field from x = 0 to x = Y is variously referred to as the "sheath," "cathode fall," and/or "positive ion space charge." Equations (10) - (13) are valid only if Y < W, where W is the plate spacing, or anode position. These conditions are referred to as subsaturated (Y < W) or saturated (Y = W).

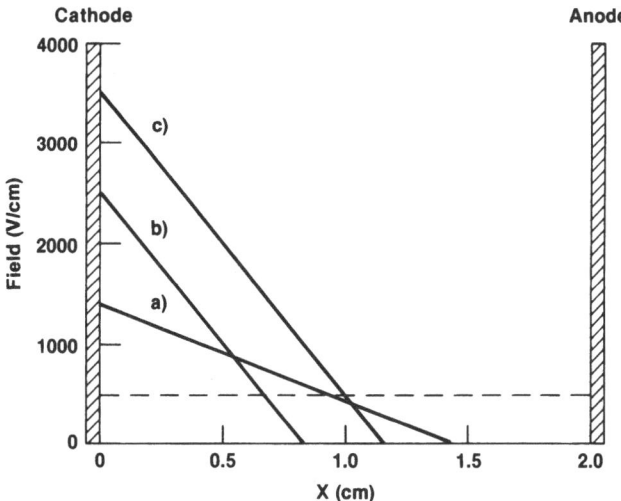

Fig. 2. Electric field distributions in flames, calculated with the
one-dimensional model of Lawton and Weinberg[16] for (a) r =
$10^{13} cm^{-3} sec^{-1}$, V_a = 1000 V; (b) r = $10^{14} cm^{-3} sec^{-1}$, V_a = 1000
V; and (c) r = $10^{14} cm^{-3} sec^{-1}$, V_a = 2000 V. In all cases,
an ion mobility of 20 $cm^2 V^{-1} sec^{-1}$ was assumed. The dashed
line illustrates the field that would be produced by a 1000
V potential across the 2-cm spacing in the absence of a
flame.

Conditions of supersaturation (Y > W) are treated by Lawton and
Weinberg,[16] but are more complex and are less germane to relevant
analytical conditions.

     The slope in the electric field occurs because electrons are
more effectively removed by a field than ions, owing to their greater
mobility, yielding a net positive "space charge" in regions of non-
zero field. This charge, in turn, perturbs the field through
Poisson's equation

$$-dE/dx = e(n_+ - n_-)/\varepsilon_o \tag{14}$$

(in one dimension and with our inverted sign convention for x). The
field perturbation results in an equal removal rate for ions and
electrons at steady state.

     Curves (a) - (c) in Figure 2 illustrate the effect of r and V_a
on the field in the flame. In curve (a), values or r = 10^{13}
$cm^{-3} sec^{-1}$, μ_+ = 20 $cm^2 V^{-1} sec^{-1}$, and a V_a = 1000 V have been assumed.
(this r value gives a reasonable approximation to an unseeded C_2H_2/
air flame.) Moving from curve (a) to curve (b), only r has been
changed - from 10^{13} $cm^3 sec^{-1}$. In practice, such an increase might

arise from the presence of a trace level ($\sim$ ppm) of an alkali metal,
or other easily ionized species, in the flame. The effect of this
increase in volume ionization rate is to shrink the extent of the
sheath (reduce Y), while yielding higher fields at the cathode. The
integral under each of the curves (a) and (b) must be the applied
voltage, 1000V.

Curve (c) of Figure 2 illustrates the effect of raising the
applied voltage to 2000 V while maintaining the other parameters of
curve (b). Doubling the applied voltage may be seen to increase the
sheath width by $\sqrt{2}$ as indicated in equation[11]. This is an important result for LEI, since the field must generally be made to pen-
etrate the laser interaction volume for an LEI signal to be detected,
as discussed below. However, the practical limit to raising the
applied voltage to compensate for easily ionized matrices is determined by arc-over in the region of high field at the cathode.

In addition to the velocity given LEI ions by the field in the
flame, their motion is influenced by flame convection (burnt gas
velocity) and diffusion. However, the relative importance of these
three influences is indicated by the cw "imaging" experiment in a
H/air flame,[17] illustrated in Figure 3. In the apparatus, shown in
the inset, a positive or negative high voltage applied to electrode 1
repels ions or electrons generated by the laser, 2, toward electrode
5, which is grounded through a load resistor. A 1-mm diameter wire,
6, is grounded through an equivalent load resistor, and can be translated vertically to intercept ions or electrons traveling toward
electrode 5. The curves shown in Figure 3 represent LEI signal from
the translated wire as a function of wire position.[17] At high
voltage ($\pm$2000 V) the laser-produced ions and electrons are clearly
imaged at the height of the laser beam (19 mm), showing little diffusive broadening. At low voltage ($\pm$250 V), the flame velocity
translates both images, and diffusive effects broaden both images.

A computer program has been written to simulate the transport of
ions and electrons generated in a flame by pulsed LEI.[18] The
program performs a numerical solution - in one dimension - of the
incomplete (approximate) continuity equations

$$\frac{dn}{dt}_+ = \mu_+ \frac{d}{dx}(n_+ E) + D_+ \frac{d^2 n_+}{dx^2} \qquad (15)$$

$$\frac{dn}{dt}_- = -\mu_- \frac{d}{dx}(n_- E) + D_- \frac{d^2 n_-}{dx^2} \qquad (16)$$

where the diffusion coefficients for ions and electrons, D_+ and D_- are
related to the corresponding mobilities through[16]

$$D_+ = \mu_+(kT/e), \text{ and} \qquad (17)$$

$$D_- = \mu_-(kT/e). \qquad (18)$$

Equations (15) and (16) neglect the effects of convection and re-combination. The program begins from initial conditions of a linearly decreasing electric field and equivalent, narrow triangular distributions of a specified number of ions and electrons about a specified laser position. The program can be used to display a "movie" of the time evolution of the electron and ion distributions. A representative "frame" of such a movie is shown in Figure 4. The assumed electric field is plotted as well as the ion and electron distributions 160 nsec after the laser pulse. The ions have barely begun to move toward decreasing x from the initial triangular distri-bution about x = 0.3 cm. Both the drift (field induced translation) and diffusion of the electron distribution may easily be seen. The fact that the electric field decreases in the direction of motion of the electrons causes a bunching effect which counteracts diffusion to some extent.

The computer program also allows the ion and electron distri-butions to perturb the originally assumed electric field through Poisson's equation [equation (14)]. The perturbed field shown in

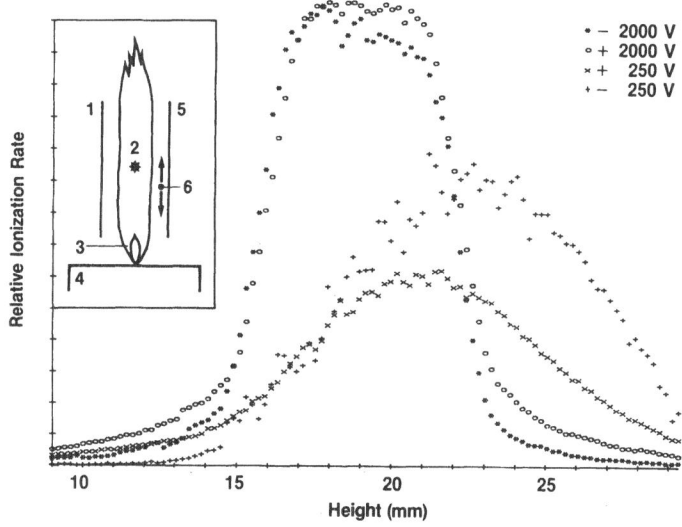

Fig. 3. Spatial images of ions at applied voltages of +250 V and +2000 V, and electrons at −250 V and −2000 V, using the imaging rod electrode configuration shown in the inset[17]: 1 = high voltage electrode; 2 = laser position; 3 = reaction zone; 4 = burner head; 5 = low voltage electrode; 6 = translating imaging (signal) electrode at the same potential as 5.

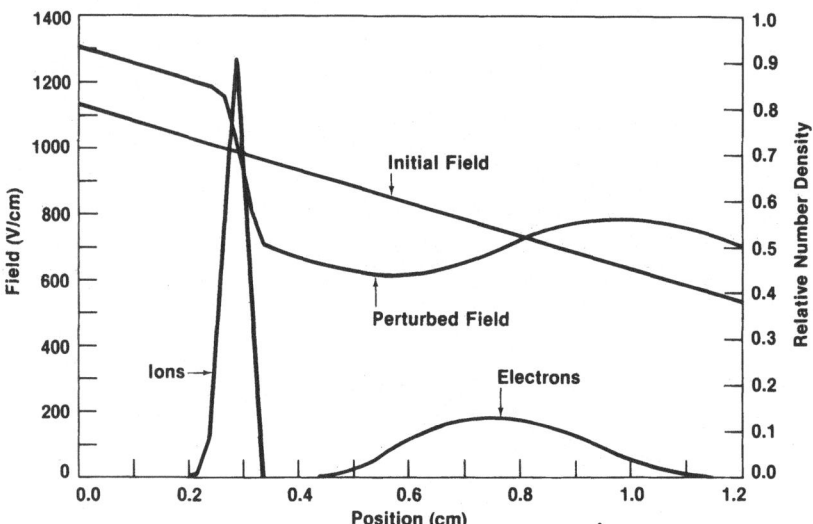

Fig. 4. Computed ion and electron distributions 160 nsec after the
         formation of 10[10] ions and electrons in a narrow distri-
         bution about X = 0.03 cm. Both the assumed initial field,
         and the field as perturbed by the charges are shown.
         Minimal field perturbation occurs at most concentrations of
         interest in LEI.

Figure 4 is an exaggerated case corresponding to high analyte
concentration.

Flame LEI Signal Currents

     The cw imaging experiment described previously (Figure 3) would
seem to imply that the LEI signal current in the detection circuit
arises from charge arrival and neutralization at the electrodes.
Pulsed LEI signals, however, are entirely inconsistent with this
interpretation, because signal generation begins before charge
arrival. The concept of induced charges and currents must thus be
invoked.

     A simple "point-charge" description has been derived[19] for the
induced LEI current for plane parallel electrodes, a uniform flame,
and a sub-saturated field. Diffusion, convection, and recombination
are ignored, and only the drift of positive and negative point
charges in the field is considered. The current i induced in the
external circuit following deposition of ion and electron charges q
and -q by a laser at position x_L (measured from the cathode) is given
by

$$i = i_+ + i_-$$ (19)

$$i_- = \dot{q}(Y - x_L)Y^{-1}\tau_-^{-1}\exp(-t/\tau_-) \tag{20}$$

$$i_+ = \begin{cases} q(Y - x_L)Y^{-1}\tau_+^{-1}\exp(t/\tau_+), & 0 \leq t \leq t_a \\ 0 & , t > t_a \end{cases} \tag{21}$$

$$\tau_- \equiv [\mu_- \delta\sqrt{r}]^{-1}, \quad \tau_+ \equiv [\mu_+ \delta\sqrt{r}]^{-1} \tag{22}$$

$$t_a \equiv \tau_+ \ln[Y/(Y - x_L)] \tag{23}$$

$$x_L < Y, \quad Y \leq W \tag{24}$$

where i_- and i_+ are the electron and ion components of the induced current, W is the electrode spacing (cm), and t_a is the ion arrival time at the cathode.

The electron component may be seen from equation (20) to have its maximum at t = 0, and to approach zero exponentially with a time constant (τ_-) governed by the field slope [equations (10) – (12)] and electron mobility. This behavior results from the fact that the electrons move in the direction of decreasing field, with the maximum velocity and induced current at the point of origin (x_L), and approach zero velocity as they approach x = Y. The ions, on the other hand, move in the direction of increasing field, and hence travel with increasing velocity and induced current, until they encounter the cathode, abruptly terminating the current source. This behavior yields the truncated, positive exponential ion behavior of equation[21].

Because the characteristics times τ_- and τ_+ vary inversely with mobility [equation (22)], electron and ion pulses occur in vastly different time and intensity domains. In analytical LEI, gated detection is normally restricted to 1 ÷ 2 μsec, which includes the electron pulse and very little of the ion pulse. The ion pulse may be used to measure mobilities through the arrival time, t_a.[20]

The restrictions of equation (24) apply to all of equations (19) – (23). The restriction $Y \leq W$ results from the assumption of sub-saturation fields in the derivations. Physically, the requirement that $x_L < Y$ means that the LEI even must occur in a region of non-zero field in order for the field-produced charge motion to induce current in the external circuit in predictable synchronism with the laser. Continuous wave LEI is more forgiving in this regard, since charged species can be moved into the sheath by convection or diffusion and still be detected within the synchronization window, for chopping rates < 500 Hz.[21]

ANALYTICAL DEVELOPMENT

Instrumentation

A generalized block diagram of a pulsed LEI spectrometer is shown in Figure 5. The light from one or two tunable laser systems is directed into a flame supported on a commercial atomic absorption (pre-mix, laminar flow) burner, into which sample and calibration solutions may be aspirated. A high voltage is applied across the flame, requiring at least two electrodes, of which the burner head itself may be one.

A simple resistor-capacitor network is used to shunt the dc background current to ground through a load resistor while directing the LEI signal pulses to a pulse amplifier. Amplified signal pulses are time-filtered and processed by a boxcar averager synchronized to the laser, whose analog output may be recorded on a chart recorder or other data handling system.

Although cw lasers have some advantages for analytical LEI,[21] pulsed lasers presently have more general applicability because of their superior wavelength range and UV generation capability. The pulsed system currently in use at NBS is a frequency-doubled or tripled Nd:YAG laser simultaneously pumping two tunable dye lasers, of which one may be frequency doubled into the UV. The choice of this system is discussed further below.

Any pulsed laser system will produce a certain amount of radio-frequency interference (RFI) from which it is difficult to shield the open electrode system used in LEI. Though special care in shielding and grounding are worthwhile, the single most useful device to minimize the effect of RFI has been a high-gain current-to-voltage

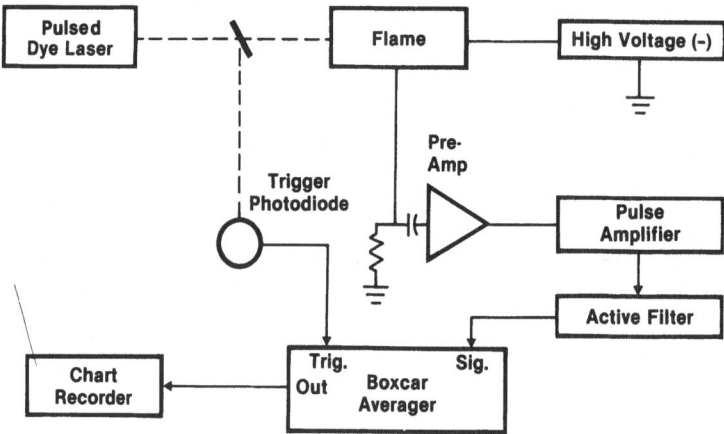

Fig. 5. Block diagram of a typical LEI spectrometer using a pulsed laser.

preamplifier close-coupled to the low-voltage electrode.[11]   The 10^6
V/A preamp most commonly used in analytical LEI studies to date[22]
has a bandwidth of $\sim$ 1MHz. Thus, subtle changes in the submicro-
second electron pulse shape with changing experimental conditions, as
suggested by equation (20), are normally lost due to the relatively
slow response of the preamplifier.

Almost any conceivable electrode scheme will work, to some
extent, for LEI. End-on views of several electrode systems are
illustrated in Figure 6, in the order of development.[10] The evolution in electrode design has been in response to a matrix interference discussed below (see Accuracy). All of the electrodes have been
of simple design, beginning with 1 mm (0.040 inch) diameter tungsten
welding rods; followed by sheet metal (stainless steel, Al, Mo)
plates hand-sheared to size, and, finally, the 6 mm (0.25 inch)
diameter stainless steel tubing, slightly flattened in a vise. The
latter electrode (Figure 6d) is readily water-cooled, and can be
immersed in the flame without deterioration.

Holders or clamps for electrodes[23] must be properly insulated
for high voltage (> 2000 V). When the burner head is used as one
electrode, it is used as the low voltage electrode, and negative high
voltage is applied to the other electrode(s). High voltage polarity
is determined by the fact that the highest fields exist near the
cathode [equation (10) and Figure 2]. Hence, the cathode should be
the electrode closer to the laser interaction zone.

Sensitivity

It was shown earlier that every analyte atom in the interaction
volume can be ionized during a laser pulse, given an appropriate
excitation scheme. This fact gives LEI its extraordinary sensitivity, with limits-of-detection* (LOD's) in the pg/mL to ng/mL range
for 24 elements to date. These elements are shown in italics in the
periodic table in Figure 7,[10] along with the excitation scheme used
(defined in Figure 1) and the LOD, in ng/mL. Other elements shown
are amenable to flame spectroscopy, and should yield favorable
results by LEI. Most of these elements require the use of a C_2H_2/N_2O
flame (in place of the normally used C_2H_2/air flame) in order to
provide a reasonable free atom fraction. LEI experience with this
flame is limited to date to the Al result shown in Figure 7.
Elements omitted totally from Figure 7 are those not commonly studied
by any flame spectrometric method.

---

* The LOD convention used here is the one prevalent in analytical
  atomic spectrometry; i.e., the analyte concentration for which the
  signal is three times the rms noise of the blank.

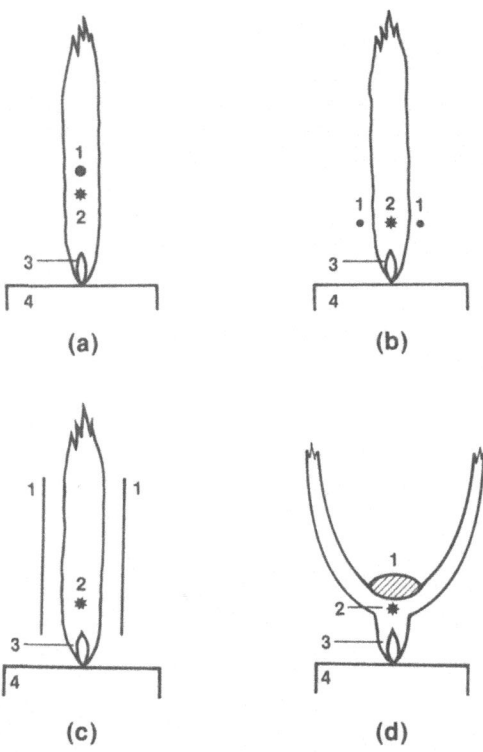

Fig. 6.   End-on representation of several electrode configurations
          used for LEI[10]: (a) immersed rod; (b) split cathode, rods;
          (c) split cathode, plates; (d) immersed, water-cooled
          cathode. In all cases, 1 is (are) the cathode(s), 2 the
          laser interaction volume, 3 the flame reaction zone, and 4
          the burner head, also used as the anode.

     The development of stepwise excitation LEI[13,14] has been
crucial to the extension of sub-ng/mL LOD's to elements with ionization potentials > 7.5 eV. The LOD of Cu, for instance, was improved by 3 orders of magnitude.[14] Stepwise excitation is particularly simple to implement with a pair of dye heads pumped by a
single N_2, Nd:YAG, or excimer laser, though synchronized flashlamp-
pumped dye lasers may be used as well. The relatively long pulse
length of flashlamp-pumped dye lasers ($\tau_L \approx 10^{-6}$ sec) would appear to
yield a sensitivity advantage over the above-mentioned laser-pumped
dye lasers ($\tau_t \approx 10^{-8}$ sec) through the time dependence of ion production
shown in equation 6. However, if stepwise excitation is used to
populate high-lying levels such that $k_m F \gg 10^8$ sec^{-1}, then the
function in equation 6 is asymptotically approaching total ionization, and the difference in pulse length results in a trivial
difference in ion yield. Thus, the use of a single pump laser to
power two dye heads is preferred for its simplicity of operation.

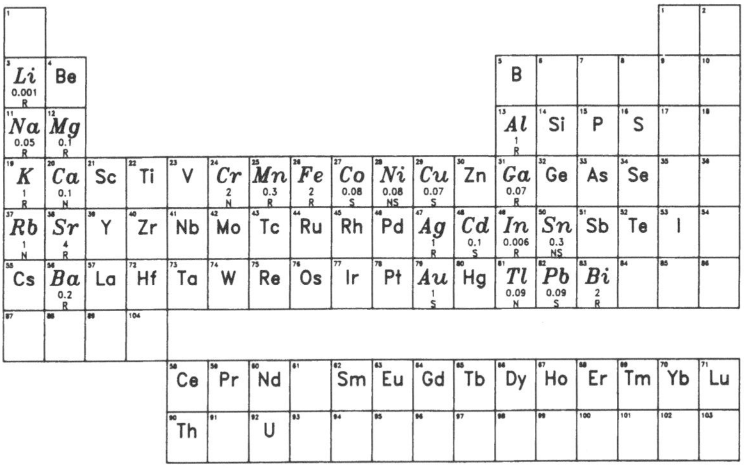

Li 0.001 R	Be											B					
Na 0.05 R	Mg 0.1 R											Al 1 R	Si	P	S		
K 1 R	Ca 0.1 R	Sc	Ti	V	Cr 2 N	Mn 0.3 R	Fe 2 R	Co 0.08 S	Ni 0.08 NS	Cu 0.07 S	Zn	Ga 0.07 R	Ge	As	Se		
Rb 1 N	Sr 4 R	Y	Zr	Nb	Mo	Tc	Ru	Rh	Pd	Ag 1 R	Cd 0.006 S	In 0.006 R	Sn 0.3 NS	Sb	Te	I	
Cs	Ba 0.2 R	La	Hf	Ta	W	Re	Os	Ir	Pt	Au 1 S	Hg	Tl 0.09 N	Pb 0.09 S	Bi 2 R			

| Ce | Pr | Nd | | Sm | Eu | Gd | Tb | Dy | Ho | Er | Tm | Yb | Lu |
| Th | | U | | | | | | | | | | | |

Fig. 7. LEI periodic chart of the elements,[10] indicating
experimental limit-of-detection (ng/mL) and excitation mode
[see Figure 1] for elements observed to date (italics).
Other elements shown are expected to yield LEI signals in
flames; omitted elements are not amenable to flame
spectrometry.

Theoretical calculations of the ultimate sensitivity which may
be expected from LEI in flames[24] agree well with the best LOD
shown in Figure 7, i.e., Li at 1 pg/mL, which corresponds to $\sim 10^5$
free atoms per cm^3 in the flame. The "ideal" calculation assumes a
single limiting noise, specifically, the "shot noise" associated with
the natural flame background current.

The LOD's shown in Figure 7 were obtained using low concentration solutions of appropriate purified salts in distilled water.
Under more practical conditions of multicomponent samples with high
concentrations of matrix species, several factors may occur to degrade sensitivity. When easily ionized species are present in a
sample at high (> 1 ppm) concentrations, the flame background current
will increase, and the associated shot noise will increase as the
square root of the background current, raising LOD's accordingly.

Another practical limitation to sensitivity may be aggravated by
the presence of particular species in sample matrices, but may also
account for some of the discrepancy between theory and experiment for
ideal solutions as well. This factor is the generation of unwanted
ions of matrix or natural flame species by the laser. Such non-
resonant ionization may occur by multiphoton ionization and/or direct
photoionization of thermally excited species in the flame. This
noise source carries with it the effect of pulse-to-pulse variation
in the laser output, which may be on the order of 5 percent for may
pulsed lasers.

Accuracy

Quantitation with LEI may be achieved in several ways. The simplest method is to tune the laser to the maximum LEI signal from the element of interest, and sequentially aspirate the sample or calibration solution and the corresponding blank. Differences in instrument response between a solution and the corresponding blank are then taken to represent the LEI signal. The concentration of analyte in the sample may be inferred from a calibration curve obtained with known standard solutions. Subtraction of the instrument response to the blank from the instrument response to the sample ostensibly corrects for background instrument response arising from sources other than the analyte. This approach is reasonably valid if the background response is equivalent for all of the solutions involved. In LEI, such a case would arise if background signal was produced only by laser multiphoton ionization of indigenous flame species.

If a sample contains a species which is subject to nonresonant ionization by the laser (wavelength-independent over a few nm), the above method of quantitation is subject to error (unless the sample blank contains the same concentration of the offending matrix species as the sample). In such cases, the wavelength tunability of the laser may be used to achieve accurate background correction. The use of a blank in the original procedure is simply replaced here by a measurement of instrument response to the sample solution with the laser tuned off of the atomic transition by several linewidths. The sample thus becomes its own, perfectly "matrix-matched" blank.

A more difficult source of error to cope with arises from spectral coincidences of the analyte transition with a transition of an atomic or molecular matrix species. Off-resonance background correction is invalid in such cases since tuning the laser affects the background response as well as the analyte response. However, stepwise excitation LEI is able to provide accurate results. An example of such an application of stepwise LEI is shown in Figure 8.[14] Spectral scan (a) was taken with 100 ng.mL Co aspirated into the flame, one laser tuned to the 252.136 nm Co resonance line, and the second laser scanned across a Co excited state transition at 591.680 nm. The base-line in (a) - well off the excited-state resonance - represents the single-color (252.136 nm) Co LEI response, plus a small amount of multiphoton ionization background. In curve (b), 100 ng/mL In is present in addition to the 100 ng/mL Co which was present in (a). A 252.137 nm transition in In represents a near-perfect spectral coincidence with the 252.136 nm Co transition, within the $\sim$ 0.005 nm laser bandwidth. The off-resonance baseline signal in (b) represents single color LEI response from <u>both</u> Co and In, in addition to the small multiphoton ionization background. The potential error in a single color LEI determination of Co in the presence of In can be plainly seen. However, the peak-to-baseline response of Co as a result of using the second laser may be seen [by comparison of (a) and (b)] to be unaffected by the presence of In. Thus, two-color (stepwise) LEI benefits selectivity greatly because of the vanish-ingly small probability of any species having sequential transitions (sharing a common energy level) that <u>both</u> coincide with sequential transitions of another species. A small probability exists that the two analyte transitions could coincide with transitions of two <u>diff</u> <u>erent</u> matrix species, but this probability is reduced by the fact that the excited state stepwise transition is normally visible light, a spectral region with relatively few resonance transitions.

One of the original matrix interference effects of LEI[11] has been greatly alleviated by the evolution in electrode design. In pulsed LEI, both ion and electron components of the signal depend on the electric field distribution in the flame. This distribution, in turn, depends on the volume ionization rate in the flame, as illus-trated in equations (10) – (12) and Figure 2 for plane parallel plate electrodes. For rod electrodes [Figure 6 (a) and (b)], the electric field gradient is even greater than for plate electrodes, for a given volume ionization rate. The electric field thus drops to zero closer to a rod cathode than a plate cathode for a given r and applied voltage. Since the flame ionization rate may be significantly in-creased by easily ionized matrices, LEI response is thus a function of electrode shape, sample matrix, and spacing between the laser and the cathode(s).

For the split-cathode arrangement [Figure 6 (b) and (c)], Green et al.,[25,26] studied LEI response as a function of sodium matrix

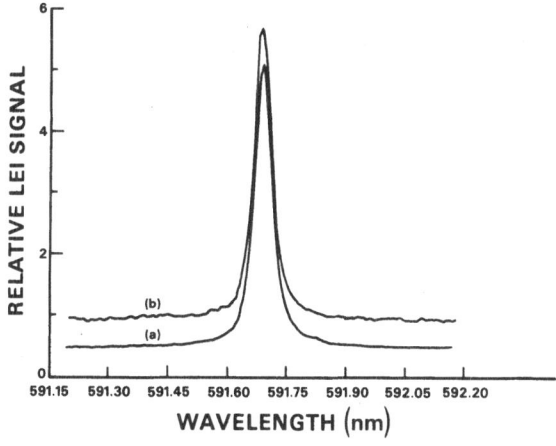

Fig. 8. Stepwise LEI wavelength scan over the Co second step transition at 591.680 nm for 100 ng/mL Co. (a) Without In present. (b) With 100 ng/mL In present. The first step Co transition at 252.136 nm overlaps the In transition at 252.137 nm.[14]

concentration for several rod diameters and for plate electrodes.
For all electrodes used, it was possible to introduce sufficient
sodium to completely eliminate the LEI signal, though the necessary
concentration of sodium was a factor of ∿ 20 higher with flat electrodes than with 1-mm diameter rods.[25]

Since the loss of LEI signal corresponds to the collapse of the
sheath (or non-zero field region) toward the cathode with increasing
ionization rate, the laser interaction volume should be as close to
the cathode as possible to further forestall loss of signal. This
last improvement was achieved with the water-cooled electrode of
Turk,[27] shown in Figure 6 (d). The slight flattening of the stainless steel tubing was intended to retain the advantage of flat plates
over rods, but with the large (6 mm) diameter of the tube, the effect
of flattening is not very important.

Figure 9 shows a comparison of LEI signal recovery for flat
plate electrodes and the immersed electrode.[27] The tolerance to
volume ionization rate for the immersed electrode is limited by
arc-over at the cathode instead of loss of LEI signal, and is quite
adequate for practical applications.

The increase in signal recovery at the end of the recovery curve
for the immersed electrode [Figure 9 (b)] is an artifact of the
separate electron and ion signals[19] [equations (19) – (24)]. Under
normal conditions, the LEI detection gate is set to bracket the
electron pulse, largely ignoring the much longer ion pulse. However,

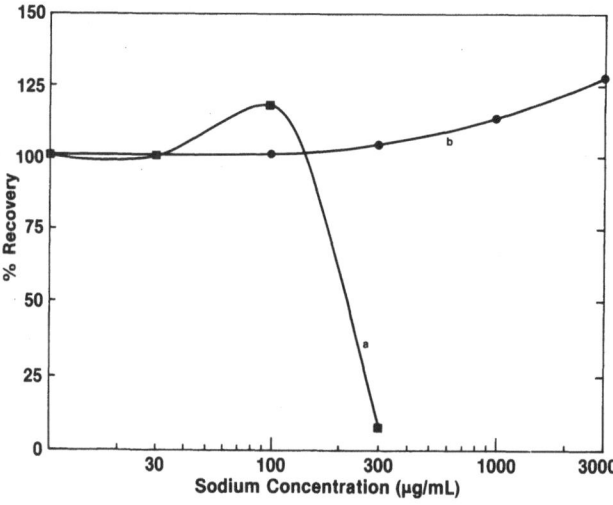

Fig. 9. LEI signal recovery for 50μg/mL Fe as a function of Na
         matrix concentration for – 1500 V applied to (a) an external
         split-plate cathode, and (b) a water-cooled immersed
         cathode.[26]

at high volume ionization rate, very high fields exist near the cathode, "pulling" a larger fraction of the ion current pulse into the time-gated detection window with the electron pulse. The increase in signal recovery just preceding signal suppression for the plate electrodes [Figure 9 (a)] may have a similar origin, but this case would require two-dimensional mathematical modeling because of the 3-electrode geometry.

The process of experimentally verifying the accuracy of LEI utilizing NBS standard reference materials is resently underway, with some results published.[14] Preliminary results indicate that high accuracy may be as important a characteristic of flame LEI as is high sensitivity.

THE OUTLOOK FOR ANALYTICAL OPTOGALVANIC SPECTROSCOPY

Optogalvanic spectroscopy in flames, or LEI, may be seen to be well on its way to becoming a legitimate analytical tool. Remaining goals in the development of LEI are:

1) the accumulation of experience with complex samples;
2) expanding the coverage of the periodic table; and,
3) the establishment of theoretical models and experimental protocols to assure high levels of accuracy.

Progress is already being made toward these goals, and they all appear to be well within sight.

Although analytical development of the optogalvanic effect has been restricted to flames, the future may well see applications in other analytical plasmas as well. Several atmospheric pressure electrical plasmas, as well as the low pressure glow discharge, would be appropriate for feasibility experiments

Acknowledgements

Long and fruitful collaboration with J. R. DeVoe, G. C. Turk, P. K. Schenck, R. B. Green, G. J. Havrilla, T. C. O'Haver, W. G. Mallard, and K. C. Smyth is most gratefully acknowledged

REFERENCES

1. R. B. Green, R. A. Keller, G. G. Luther, P. K. Schenck, and J. C. Travis, Galvanic Detection of Optical Absorptions in a Gas Discharge, Appl.Phys.Lett., 29:727 (1976).
2. R. B. Green, R. A. Keller, P. K. Schenck, J. C. Travis, and G. G. Luther, Opto-Galvanic Detection of Species in Flames, J.Am.Chem.Soc. 98:8517 (1976).

3. P. K. Schenck and J. W. Hastie, Optogalvanic Spectroscopy –
    Applications to Combustion Systems, Opt.Eng., 30:522 (1981).
4. J. C. Travis and J. R. DeVoe, The Optogalvanic Effect, in:
    "Lasers in Chemical Analysis," G. M. Hieftje, J. C. Travis,
    and F. E. Lytle, eds., The Humana Press, Clifton, N.J. (1981).
5. J. E. M. Goldsmith and J. E. Lawler, Optogalvanic Spectroscopy,
    Contemp.Phys., 22:235 (1981).
6. F. M. Penning, Demonstratie van een Nieuw Photoelectrisch
    Effect, Physics 8:137 (1928).
7. P. D. Foote and F. L. Mohler, Photo Electric Ionization of
    Caesium Vapor, Phys.Rev. 26:195 (1925).
8. C. Staneiulescu, R. C. Bobulescu, A. Surmeian, D. Popescu, I.
    Popescu, and C. B. Collins, Optical Impedance Spectroscopy,
    Appl.Phys.Lett., 37:888 (1980).
9. K. C. Smyth, U. S. National Bureau of Standards, Washington, D.
    C. 20234, private communication.
10. J. C. Travis, G. C. Turk, and R. B. Green, Laser-Enhanced
    Ionization Spectrometry, Anal.Chem., 54:1006A (1982).
11. G. C. Turk, J. C. Travis, J. R. DeVoe, and T. C. O'Haver,
    Analytical Flame Spectrometry with Laser Enhanced Ionization,
    Anal.Chem., 50:817 (1978).
12. K. C. Smyth, P. K. Schenck, and W. G. Mallard, What Really Does
    Happen to Electronically Excited Atoms in Flames?, in: "Laser
    Probes of Combustion Chemistry," D. R. Crosley, ed., ACS
    Symposium Series 134, American Chemical Society, Washington,
    D. C. (1980).
13. G. C. Turk, W. G. Mallard, P. K. Schenck, and K. C. Smyth,
    Improved Sensitivity for Laser Enhanced Ionization Spec-
    trometry in Flames Using Stepwise Excitation, Anal.Chem.,
    51:2408 (1979).
14. G. C. Turk, J. R. DeVoe, and J. C. Travis, Stepwise Excitation
    Laser Enhanced Ionization Spectrometry, Anal.Chem., 54:643
    (1982).
15. P. K. Schenck, J. C. Travis, G. C. Turk, and T. C. O'Haver,
    Laser Enhanced Ionization Flame Velocimeter, Appl.Spectrosc.,
    36:168 (1982).
16. J. Lawton and F. J. Weinberg, "Electrical Aspects of Combus-
    tion," Clarendon Press, Oxford (1969).
17. P. K. Schenck, J. C. Travis, G. C. Turk, and T. C. O'Haver,
    Collection of Ions Produced by Continuous Wave Laser-Enhanced
    Ionization in a Hydrogen/Air Flame, J.Phys.Chem., 85:2547
    (1981).
18. P. K. Schenck and J. C. Travis, U. S. National Bureau of
    Standards, Washington, D. C. 20234, unpublished.
19. G. C. Turk, G. J. Havrilla, J. C. Travis, and P. K. Schenck, in
    preparation.
20. W. G. Mallard and K. C. Smyth, Mobility Measurements of Atomic
    Ions in Flames Using Laser Enhanced Ionization, Comb.and Flame
    44:61 (1982).

21. G. J. Havrilla, S. J. Weeks, and J. C. Travis, Continuous Wave
     Excitation in Laser Enhanced Ionization Spectrometry, Anal.
     Chem., 54:2566 (1982).

22. G. J. Havrilla and R. B. Green, Pre-amplifier for Laser Enhanced
     Ionization Spectrometry, Chem.Biomed.Environ,Instrum., 11:273
     (1981).

23. G. J. Havrilla and R. B. Green, Electrode Positioner for Laser
     Enhanced Ionization Spectrometry, Anal.Chem., 53:134 (1981).

24. J. C. Travis Limits to Sensitivity in Laser Enhanced Ionization,
     J.Chem.Ed., 11:909 (1982).

25. R. B. Green, G. J. Havrilla, and T. O. Trask, Laser Enhanced
     Ionization Spectrometry: Characterization of Electrical Inter-
     ferences, Appl.Spectrosc., 34:561 (1980).

26. G. J. Havrilla and R. B. Green, Evaluation of Plate Electrodes
     for Laser Enhanced Ionization Spectrometry, Anal.Chem.,
     52:2376 (1980).

27. G. C. Turk, Reduction of Matrix Ionization Interference in Laser
     Enhanced Ionization Spectrometry, Anal.Chem., 53:1187 (1981).

APPLICATION OF LASERS TO TRACE ANALYSIS IN THE

ATMOSPHERE AND EXPERIMENTS WITH SINGLE ATOMS

H. Walther

Sektion Physik, Universität Müchen and
Max-Planck-Institut für Quantenoptik
Garching, Fed.Rep.Germany

LASER INVESTIGATIONS IN THE ATMOSPHERE

Introduction, Survey of Methods

The laser is an ideal instrument for determining properties of the atmosphere by absorption and scattering processes: its high spectral density and low divergence are especially useful for such measurements. Lasers with continuous and discrete tunability can be used to measure specific components selectively. In the following a brief survey of methods used to date is given and new results reported.

Scattering of laser light in the low atmosphere is dominated by Mie scattering caused by aerosols, clouds, dust and other particulates. Rayleigh scattering from the molecular constituents of the atmosphere is two orders of magnitude lower. It becomes dominant above heights of 30 km, where virtually no aerosols exist. It is therefore possible to derive from the scattering intensity at these heights information about the gas density. From these, one can determine the pressure and temparature distributions; even seasonal variations of these parameters were studied in RADAR-like experiments.[1,2]

For the analysis of gases in the lower atmosphere, fluorescence or Raman scattering can be measured. Absorption measurements yield the highest sensitivity and are very simple, but give concentrations only integrated along the path of the light.[3] To cover a larger area by the laser bean, mirrors can be used. A very simple setup uses topographic targets for reflection of the light back to its source for detection. By the use of mirror arrays, absorption measurement, like tomography, can yield the spatial distribution of gas constituents.[4]

Normally, absorption measurements can be performed with low-power lasers (e.g. diode lasers). Even arc lamps with spectral filtering can be applied. This setup was very successful in detecting SO_2, N_2O, CH_2O, O_3, NO_2 and NO_3 with very high sensitivity. [5,6,7]

A remarkable improvement of the detectivity in absorption measurements becomes possible by a heterodyne technique, where a tunable laser is used as local oscillator and a photodetector as mixer.[8] The signal at the intermediate frequency is amplified by a suitable amplifier of narrow bandwidth. This technique (most advantageous in the infrared spectral range) yields an increase in sensitivity of several orders of magnitude, allowing appreciable lengthening of the absorption paths. With heterodyne detection, fast detectors used at room temperature (like pyroelectric detectors) are equally sensitive as cooled infrared detectors.

For direct detection of the radiation emitted by a pollutant, the heterodyne method is very useful, too. Of course, it can work only if the temperature of the gas being measured is higher than the ambient temperature (as in the exhaust of a chimney). The heterodyne method is already being used in satellite - or balloon-borne experiments for detecting atmospheric gases (e.g. O_3) by absorption with the skylight serving as light source.[9]

Absorption measurements, although of highest sensitivity, nevertheless have the disadvantage of delivering information about the concentration distribution only under certain restricting conditions. Studies making use of atmospheric scattering are free from this disadvantage since time-dependent observation of the backscattered light from a pulsed laser allows a spatial resolution (as in RADAR). Since light replaces the radio waves, these methods are known under the name of LIDAR. Table 1 summarizes the properties of the atmosphere that can be measured by means of scattering processes.

The relative intensity of the scattering is given in Table 2. From these values it is obvious that Raman scattering is useless for the detection of gases at lower concentrations. For most pollutants, the ratio N_p/N_a is of the order of 10^{-6}. It follows that the Raman signal of the pollutant is 9 orders of magnitude lower than the Rayleigh signal, to which all gases contribute.

The use of fluctuations of the refractive index for anemometry and the study of turbulences has been investigated by several authors. An exact analysis of these methods is given in[10]. For anemometry for example two parallel beams of a He-Ne laser are transmitted through the atmosphere, at a distance of 0.5 to 1 m. The

Table 1. Properties of the Atmosphere that can be Studied by Lasers

Process	Information content
Rayleigh scattering	total density distribution, temperature and pressure distributions
Mie scattering	density and size of aerosols, clouds, dust, smog
Raman scattering	molecules, temperature
Resonance scattering	atoms, molecules, temperature
Resonance scattering	atoms, molecules, temperature
Resonance absorption	atoms, molecules, temperature
Fluctuations of refractive index	turbulence, wind velocity

Table 2   Comparison of the Relative Intensity of Backscattered Signals

Scattering Process	Relative signal	Remarks
Raleigh scattering	1	
Mie scattering	80	at a visibility distance of 5 km
Raman scattering	10^{-3} Np/Na	Np, Na = densities of pollutant and of total atmosphere resp.
Fluorescence scattering	80	at a scattering cross-section of 10^{-16} cm^2 (electronic transitions of molecules and the assumption that 1% of the absorbed energy is reemitted as radiation and the concentration is 0.3 ppm

movement of the air causes a correlation between the intensity fluctuations of the two beams; the component of the wind velocity lying in the plane of the beams and normal to them can be evaluated from the correlation time and the distance of the beams.

Anemometry can also be performed by measuring the frequency shift caused by the Doppler effect. The frequency shift of the

backscattered (Mie scattering) signal versus the laser frequency
allows determination of the velocity component parallel to the laser
beam. A cw laser is used. Information about the wind velocity can
also be derived from photon correlation measurements of the backscattered light.[10]

Reference [1,2] gives a survey of the various results obtained
by LIDAR methods in connection with Mie and Rayleigh scattering. As
an example, we briefly report the results of Svanberg [11], who
observed the smoke plume from a chimney by Mie scattering (Figure 1).
A three-dimensional picture of the distribution is obtained by
directing the laser beam in different directions. The use of Mie
scattering facilitates detection of the plume, but it is difficult to
evaluate from the scattered signal the mass concentration of the
particulates. This is not the case in measurements of gas constituents, as discussed below.

In principle, the observation of fluorescence scattering for
detecting certain geaseous constituents of the atmosphere is useful,
too. In the lower atmosphere problems arise from collisional quenching of the fluorescence. In general, only 1% of the atoms can
radiate their excitation energy, whereas the majority transfers this
energy to the molecules of the air by collisions. The value of the
scattering cross section of Table 2 relates to electronic transitions
of molecules, it is 4 to 5 orders of magnitude lower for vibrational
transitions in the infrared. For electronic transitions of molecules, the fluorescence is distributed over a wide range of the
spectrum since the ground state is split up into vibrational levels.
In order to detect sufficient fluorescent light the receiver system
has to be broad-banded. This, of course, lowers the ratio between
fluorescent and background signals.

The detection of atoms in the higher atmosphere by fluorescence
radiation is, however, much easier: a LIDAR experiment of this kind
was performed by Sandford et al.,[2,13] for the first time. Na
atoms were detected in these measurements; other alkaline metals were
found in later ones. To minimize the background, the first experiments could only be conducted at night. As more powerful lasers are
now available, it is also possible to measure during the day. The
temporal variation of the atomic density can now be observed around
the clock.

Dye lasers with pulse energies of up to 100 mJ were used. The
laser is tuned to a resonance line; the fluorescence radiation is
collected by a large aperture mirror. A (generally double-
structured) layer about 15 km thick was found at an altitude of 90
km, the Na atom density being of the order of 10^3 cm^{-3}. The laser
experiments clearly established a coincidence between the sodium
concentration and the occurrence of meteorites. One can therefore be
sure that at least some of the alkaline atoms arrive in the atmos-

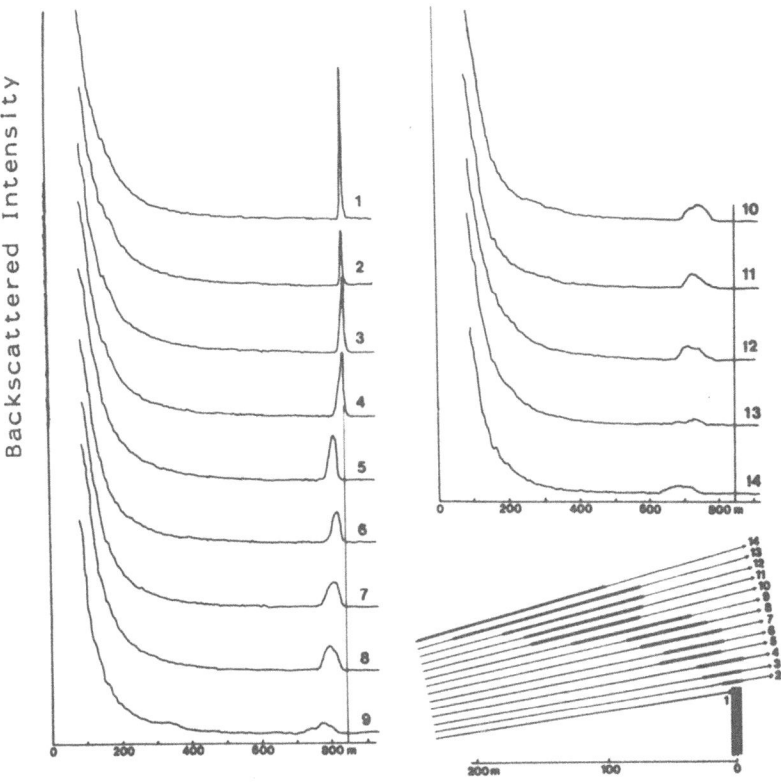

Fig. 1.   Observation of a plume by laser.   A nitrogen laser was
used.   The backscattering signal is plotted as a function of
time and also of distance R.   As the divergence of the laser
beam is low and the backscattering is isotropic, the signal
shows a 1/R dependence.   For details, see Ref.[11]

phere by evaporation of meteorites.   For the formation of atomic
sodium, the ratio of the concentrations of atomic oxygen to ozone is
essential, as the following reactions take place:

$$NaO + O \rightarrow Na^* + O_2$$

$$Na + O_3 \rightarrow NaO + O_2$$

$$Na_2 + O \rightarrow NaO + Na^*$$

The reaction of atomic sodium and atomic oxygen taking place in
a triple collision can be neglected at the low densities.   In the
first and third reactions, Na atoms are produced in the excited ^2P
state.   Probably these reactions contribute to the aurora.   Atomic
oxygen is produced by photodissociation of O_3 and O_2.   Obviously, the

O/O_3 ratio is of a value ensuring the survival of atomic sodium just
at an altitude of 90 km.

The Differential Absorption Method

As mentioned above, the observation of absorption excludes
measurement of the density distribution of a gas. This disadvantage
can be overcome by using Mie scattering (which is very strong in the
lower atmosphere) as a "mirror". The position of the "mirror" can be
determined by the time elapsing between the emission of the laser
pulse and the detection of the back-scattered signal. The dependence
on wavelength of the Mie scattering is low. It is therefore possible
to eliminate the local variation of the scattering by measurement at
two different wavelengths. Only one of these has to be absorbed by
the gas to be detected. Both wavelengths must be sufficiently close
together, to ensure that their Mie scattering is in fact equal.

This method, called "differential absorption", was first applied
by Rothe et al.,[14,15] for measuring air pollutants. Theoretical
studies[16] had shown before that it is the most sensitive of all
LIDAR methods.

In the first measurements[14,15] the apparatus was equipped with
a tunable dye laser. The concentration of NO_2 near a chemical fac-
tory was measured. With only 1 mJ of pulse energy, concentrations as
low as 0.2 ppm could be detected at distances of 4 km. The distribution of the NO_2 concentration above the plant was determined by
varying the direction of the laser beam. Connecting points of equal
concentration measured in five different directions, the map of
iso-lines shown in Figure 2 was obtained. It becomes evident from
this picture which building is the source of the pollutant.

By means of a dye laser the differential absorption method can
also be applied to SO_2 and O_3. SO_2 measurements were carried out
e.g. by Svanberg.[11] Measurements of the ozone concentration in the
stratosphere will be discussed here in the following.

After Rowland's and Molina's publication[19] of the catalytic
cycle:

$$Cl + O_3 \rightarrow ClO + O_2$$

$$ClO + O \rightarrow Cl + O_2$$

(which, by the catalytic action of Cl atoms, converts O_3 into O_2) the
ozone problem has gained much attention. From chlorinated fluoro-
methanes (for example, used in spray cans) chlorine is produced by
photodissociation:

$$CFCl_3 \xrightarrow{h\nu} CFCl_2 + Cl$$

$$CF_2Cl_2 \xrightarrow{h\nu} CF_2Cl + Cl$$

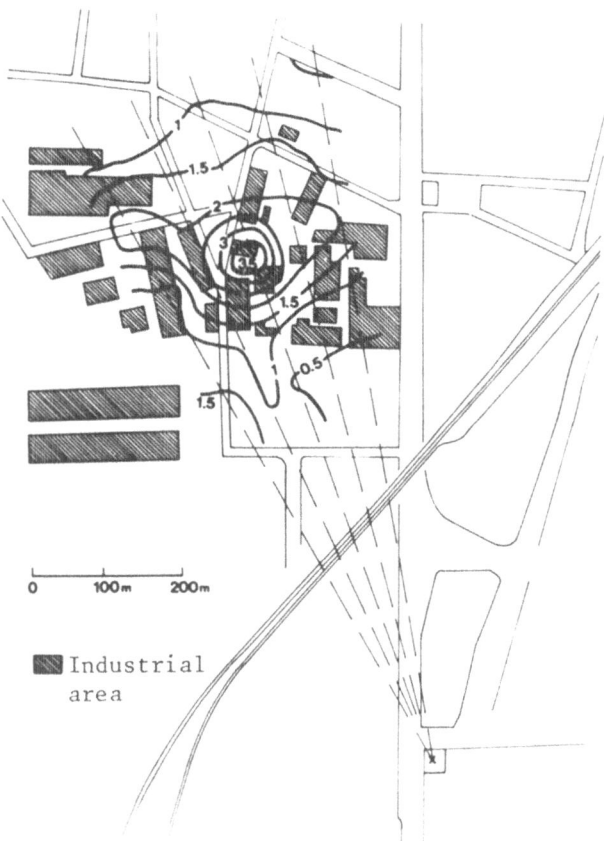

Fig. 2. Distribution of NO_2 concentration above a chemical plant. Concentrations are given in ppm. The results were obtained by averaging about 40°000 laser shots.[14,15]

A virtually no sinks exist for the chlorofluoromethanes, a steadily increasing chlorine concentration leading to the foregoing destruction of the ozone layer must be taken into account. A permanent observation of the ozone layer is therefore as necessary as it is useful.

First measurements of the ozone layer with dye lasers were performed by Gibson et al.,[12,13] and later by Megie et al.,[17] The results of Megie et al., are shown in Figure 3 as an example. The apparatus has been improved since then; today useful results about the ozone layer can be obtained from laser experiments:[18] (See also Ref. 23 for comparison.) As Mie scattering can be neglected in the upper atmosphere, Rayleigh scattering was used as "mirror" in these experiments.

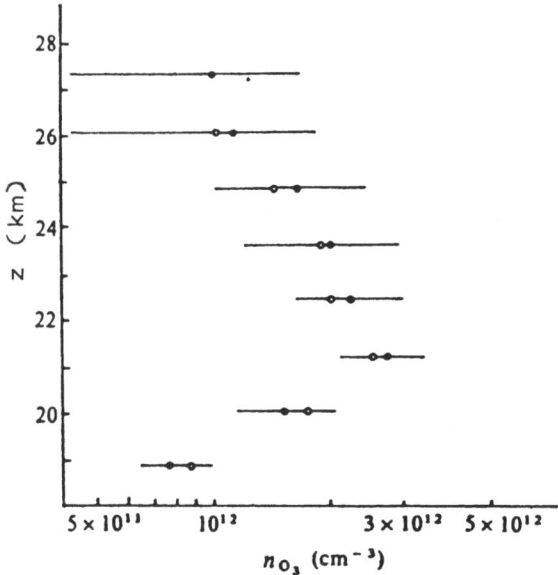

Fig. 3.   Concentration of O_3 at altitudes of 20 to 25 km taken from
          Ref[17]

In these laser measurements time intervals in the range of
several hours were required to get an accuracy of 30%. This also
holds for the measurement of Uchino et al.,[20]; they used a XeCl
laser at 308 nm without having a second frequency necessary as a
reference line for the differential-absorption technique. Instead of
this they used the data of a balloon sonde, launched the next day, as
a reference.

In Munich we have constructed a new setup for ozone measurements
in the troposphere taking into account the requirements for a high
power laser as well as for a proper reference line. We use a XeCl
laser with a pulse energy of 130 mJ and a repetition rate of up to
100 Hz. Its radiation is focussed into a high-pressure methane cell
to generate the reference line by stimulated Raman scattering.
Typical conversion rates are 15% at a pressure of 35 atm and a local
length of 125 cm. This is more than sufficient since the Raman
shifted reference line at 338 nm is absorbed significantly less by
ozone than the unshifted XeCl laser line at 308 nm.

The backscattered light is focussed by a spherical mirror (60 cm
diameter, focal length 240 cm). The two wavelengths are separated by
a dichroic beam splitter and two interference filters. The two
photomultipliers must be protected against the very strong intensities backscattered from the first few kilometers; a chopper wheel
therefore rotates in front of the multipliers, covering them during

and shortly after emission of the laser pulse. The photons back-scatterted from the stratosphere are counted, and the corresponding count-rates stored every 66.7 nsec and afterwards transferred to a computer for further data evaluation.

The major advantage of this setup is that simultaneous measurements at both wavelengths are possible, thus eliminating all problems associated with rapidly changing atmospheric conditions (turbulences). Owing to the high repetition rate and pulse energy of the XeCl laser the apparatus is expected to give ozone profiles with an accuracy of a few percent in a measuring period of 10 min.

One disadvantage of all ground-based stratospheric LIDAR measurements is the rather strong decrease of the light intensity in the first few kilometers primarily due to Mie scattering, and in the UV spectral region also due to Rayleigh scattering. The measurements with our ozone LIDAR are therefore being made from the summit of the Zugspitze in the Apl Mountains (altitude 3 km). This difference in altitude gives rise to an increase of the intensity of the strato-spheric backscattering by a factor of about three for clean air, and of twenty for hazy air.

When the differential absorption method is to be extended to a larger number of different pollutants, measurements are to be carried out in the infrared range of the spectrum. The most universal setup would be obtained by using a continuously tunable laser. As such lasers, at present, are still difficult to be used in field measurements, molecular lasers (e.g. DF, HF, CO, CO_2 and N_2O lasers) must be employed and measurement must rely on accidental coincidence of laser emission lines with absorption lines of the pollutants.[1,2,3,21,22] Several pollutants can even be detected simultaneously by a multi-line measurement or sequential measurements using different emission lines.

In the following, some results obtained by my group with a setup equipped with a multi-gas laser are reported. Details of the apparatus are published elsewhere.[3]

The first experiments to be discussed here were obtained at the cooling tower of the power station in Meppen, in cooperation with Electricité de France. The objective of these measurements was to determine the concentration of water vapour around the cooling tower in order to obtain data comparable with computational results of the distribution.

The measurements were made from two different points; the concentration was determined in different planes of varied elevation. This yielded a three-dimensional distribution of the concentration. The laser was operated with CO_2. At a pulse repetition rate of 70Hz, about 10 minutes of measurement time were necessary to obtain the distribution in one plane, as shown in Figures 4a and 4b. Figure 4 a shows the contour lines of constant concentration, while in Figure 4b the concentration is plotted on the vertical axis.

As next example, measurements of the ethylene concentration around an oil refinery near Ingolstadt are discussed. The purpose of these "in-situ" measurements was to extend the experiments to an organic gas and test the improvement of the detectivity into the ppb-range. In addition, data were to be collected that could serve as a basis for the calculation of the distribution.

The ethylene detected leaks out of the distillation plant. These measurements were also made with a CO_2 laser placed at a distance of about 500 m from the refinery. An example is shown in Figures 5a and 5b.

For environmental surveillance not only is the distribution of the pollutant interesting, but also the amount emitted per unit time. In the following, it is shown how this quantity can be determined by the differential absorption method. For this purpose the concentration distribution is measured in a vertical plane chosen normal to the wind velocity. A measurement performed in this way is shown in Figures 6a and 6b.

This distribution is compared with the results of a calculation of the distribution from a model that takes into account the "turbulent diffusion". The spatial distribution of $c(x,y,z)$ is related to the wind velocity $\bar{u}$ and the components A_{YY} and A_{ZZ} of the tensor of turbulent diffusion by the following differential equation:

$$\frac{\partial c}{\partial t} + \bar{u}\,\frac{\partial c}{\partial x} = \frac{\partial}{\partial y}\left(A_{yy}\,\frac{\partial c}{\partial y}\right) + \frac{\partial}{\partial z}\left(A_{zz}\,\frac{\partial c}{\partial t}\right)$$

It has been assumed here that the wind blows in the x-direction (see Figure 7). As the diffusion velocities are small compared with the wind velocity, the x-component of the diffusion can be neglected, which simplifies the calculations. Convection in the y- and z-directions can also be neglected. Assuming this equation, it is possible to calculate the concentration distribution in a plane (x_1,y,z).[24] Conversely, the parameters of emission can be derived from a measured distribution. In addition, the parameters for turbulent diffusion can be determined by a fit to the measured distribution. The solution is a Gaussian distribution in the y-direction; a modification of the Gaussian profile by the surface of the earth is observed in the z-direction. In general, the profile normal to the wind direction is described by a Gaussian distribution; the result of the evaluation agrees quite well with the expected distribution.

The measurements shown on Figure 6 indicates two leaks at different heights, the upper one being smaller. With a wind velocity u

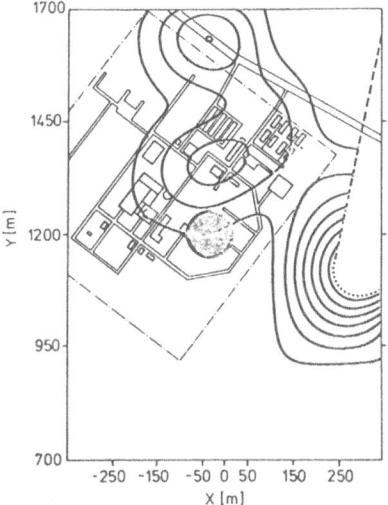

Fig. 4a.   Plan of the Meppen power plant with the cooling tower
           (round shaded area). Contour lines of the water vapour
           concentration begin with 2 Torr at the outermost line, the
           steps between the lines also being 2 Torr. The dotted line
           shows the contour of the visible plume. The plane of the
           measurement is inclined to the surface; it touches the
           surface at the position of the apparatus (x=0, y=0) and is
           at an altitude of about 100 m above the cooling tower.

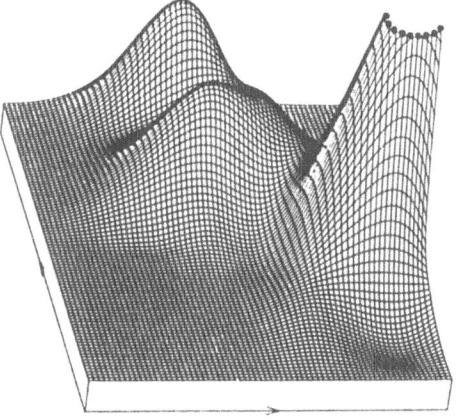

Fig. 4b.   Two-dimensional distribution of water vapour concentration
           (vertical axis) above the power plant.

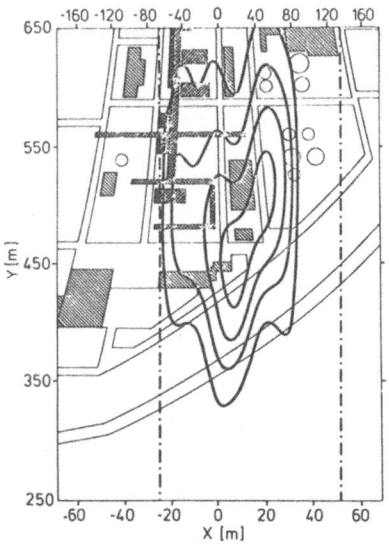

Fig. 5a. Map of the refinery and contour lines of ethylene
          concentration. The lines start with 20 ppb, the inward
          increase being 20 ppb.

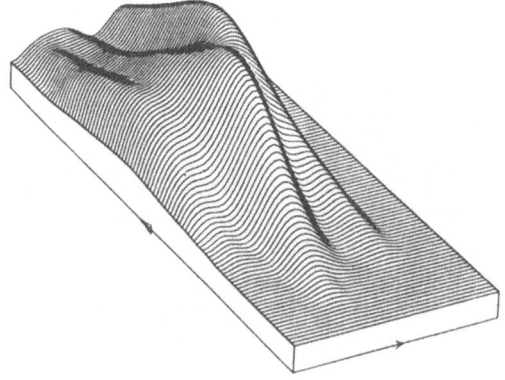

Fig. 5b. Ethylene concentration (vertical axis) above the refinery

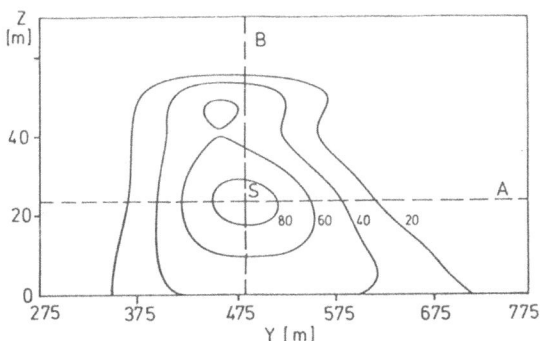

Fig. 6a. Contour lines of ethylene concentration (in ppb) in a
          vertical plane normal to the wind velocity.

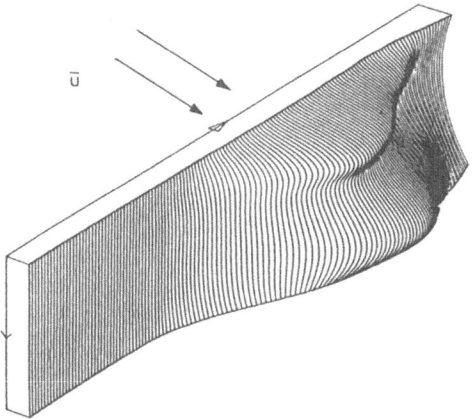

Fig. 6b. Concentration (to the right) in the plane.   Arrows
          indicate the wind.

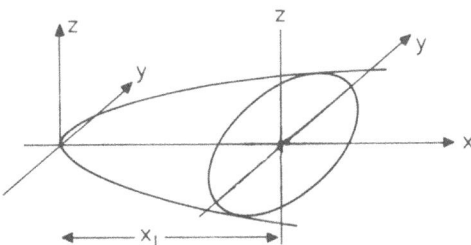

Fig. 7. To the model of gas distribution in the atmosphere.

of 3 m/sec measurement separately and assumed to be constant over the whole plane, the rate of emission of ethylene was determined to be 0.95 g/sec in the case considered here. The method used here needs to be tested further; however, it can be said that, in principle, it is useful for determining rates of emission.

EXPERIMENTS WITH SINGLE ATOMS

Fluorescence Experiments

The interaction of laser light with atomic systems has received considerable theoretical and experimental attention over the past decade. Until the advent of the laser, light sources for spectroscopy consisted of ordinary spectral lamps excited by DC or RF discharges, and produced light having a very broad spectral width, and hence very short correlation time, and having a relatively low intensity. For such fields both the experimental and theoretical results are in general well understood. However, the development of the laser afforded light sources which are sufficiently intense for an atomic (or molecular) transition to be very easily saturated. In addition, the lasers are highly monochromatic, having a coherence time much larger than typical natural lifetimes of excited atomic states, and finally, tunable, thus allowing selective excitation of particular atomic transitions. As might be expected, it is found that many new and interesting phenomena are associated with fields of this nature interacting with atomic systems.

The new physical situation requires for its theoretical analysis the use of techniques more general than those found adequate in the case of thermal fields. In the latter case, the weakness of the atom field-interaction meant that perturbative techniques were generally sufficient. These techniques were based on the assumption that the initial state of the atomic system was essentially unchanged by the interaction. However, as saturation can easily be achieved with an intense laser field, more general non-perturbative methods are required. Furthermore, for a highly coherent field, one cannot consider successive photon emission and absorption processes as being independent since it is now possible for an atomic system to undergo many such processes during the correlation time of the laser field, and hence phase memory effects are important.

Although a wide range of both theoretical and experimental problem involving laser fields have been studied, attention is confined here to just one aspect: the interaction of intense monochromatic light with atomic systems, in which it is the properties of the fluorescent light (i.e. the light scattered by the atom) that are of principal interest.

The simplest such system is also one which has attracted an enormous amount of interest: the problem of theoretically and experimentally determining the spectrum of the fluorescent light radiated by a two-level atom driven by an intense monochromatic field. This is the situation that gives rise to the AC Stark effect. For sufficiently strong fields, it is found that the spectrum of the scattered light splits into three peaks consisting of a central peak, centered at the driving field frequency with a width $\Gamma/2$ (Γ^{-1} Einstein A-coefficient) and having a height three times that of two symmetrically placed sidebands, each of width $3\Gamma/4$ and displaced from the central peak by the Rabi frequency. In addition, there appears a delta function (coherent) contribution positioned also at the driving frequency. In the limit of strong driving fields, the energy carried by this last contribution is negligible compared to the three-peak contribution. This result was first predicted by Mollow[25] and subsequently by many others, using a variety of techniques[26-37] and has been very well confirmed experimentally.[38-42]

It is not only the spectral properties of the fluorescent light that have come under investigation. The examination of the intensity correlation of the scattered field in the basic two-level atom case has also attracted much attention since it is found that the fluorescent light exhibits the property of photon antibunching.[30,31,33, 34,35] In the following, the result on photon antibunching will be briefly reviewed. For a complete review see Cresser et al.,[44] The theoretical treatment of resonance fluorescence from a two-level atom irradiated by a monochromatic light field in the low-intensity limit was first reviewed by Heitler.[45] A scattered field spectrum was predicted which was very sharply peaked around the incident field frequency. The high-intensity limit was first considered by Apanasevich,[46] who, by numerical calculations based on earlier theoretical work,[47] predicted a three-peak spectrum. Subsequently, Newstein[48] also examined the problem, but with collisional rather than radiation damping providing the relaxation mechanism. He also predicted a three-peak spectrum in the high-intensity limit, though, owing to the different damping mechanism, the widths and heights of the three peaks differed from those found later in the pure radiation damping case.

The first complete theoretical treatment in which exact expressions were obtained for the scattered field spectrum when radiation damping is present is the work of Mollow.[25] In his work, the scattering atom was driven near resonance by a monochromatic classical electric field. The atom came into equilibrium with this field through the effects of radiation damping, this being included in the theory by explicitly coupling the atom to the quantized electromagnetic field. The solution was based on deriving the optical Bloch equations for the elements of the (2x2) reduced density matrix of the atomic system. These equation were obtained from a master equation approach in the derivation of which the Markov approximation was made. The diagonal elements of this reduced density matrix are just the probabilities of the atom being found in its ground or excited state, while the off-diagonal elements essentially give the mean dipole moment of the radiating atom.

However, it is not the mean dipole moment of the atom that acts as the source of the radiated field, but rather the instananeous value of the dipole moment, i.e. its mean value plus quantum fluctuations. This is recognized in Mollow's work,[25] in which, rather than calculating the correlation function of the mean dipole moment, and hence, by a Fourier transform, the spectrum of the radiated field, it is the correlation function of the dipole moment operator that is found, so that quantum fluctuations are not averaged out. This latter correlation function is obtained from the optical Block equations by use of the quantum regression theorem.[49]

Since Mollow used a classical description of the incident field, his method was originally not believed to be a fully quantum electro-dynamic treatment, (QED), although it was later shown by Mollow[27] that this work was in fact equivalent to such a description. Many fully QED treatments have been published later; for a complete com-pilation see Ref.[44]

To characterize a light beam, study of the frequency distri-bution is usually not sufficient. Additional properties are obtained by studying the photon or intensity correlation of a light field. For usual light fields (e.g. thermal light) it is found that the intensities of the field at two neighbouring instants in time are strongly correlated, i.e. if a photo-multiplier irradiated by this field emits an electron at some instant in time, the probability is high for a second photo-emission to occur a short time later. This phenomenon is known as photon bunching, and can be explained using either a classical or quantum mechanical description of the light field.

However, it was found that for the field scattered by a single two-level atom, the intensity correlation function was of a form that showed that if a photon was detected (by a photomultiplier) at some instant in time, then the probability of detecting another photon during a short time interval following the first detection remained close to zero.

This phenomenon is the reverse of that described earlier and is known as photon antibunching.[30,31,33-35] This behavior can be explained quantum mechanically by the fact that the process of detecting a scattered photon also prepares the scattering atom in its ground state. Thus no further photons can be emitted and hence detected until the atom has sufficient time to be pumped back up to its excited state (the only state from which emission can occur) by the driving field. It should be pointed out that fields exhibiting

antibunching can be generated in other ways, i.e. by multiphoton
absorption, in which two or more photons are simultaneously absorbed,
[51,52] or else as a result of nonlinear optical effects in the
degenerate parametric process first discussed by Stoler[55] and
developed further by Paul and Brunner, Bandilla and Ritze.[50,53,54]
However, we shall be confining our attention here to antibunching in
the case of resonance fluorescence only.

The significance of antibunching is that a classical field which
exhibits this behavior does not exist. Thus the existence of photon
antibunching can be taken as a test of the validity of QED. However,
the theoretical model on which the above result is based is not
directly representative of a true experimental situation. All
experimental studies of the AC Stark effect and related phenomena
involve a beam of atoms passing perpendicularly through a laser
field. However, the antibunching effect can be washed out if a
number of atoms are simultaneously interacting with the laser field.
[33-35] Thus, ideally, to observe the phenomena, the intensity of
the atomic beam must be sufficiently low that only a single atom at a
time is passing through the field. In the real situation, however,
even for low beam intensities, there is a statistical fluctuation in
the number of atoms in the field at any time. Thus, for the purpose
of comparing theory and experiment, the above theory must be modified
to allow for fluctuations in the number of atoms in the field at any
time, and also, as it turns out, the effect of the finite transit
time of the atoms through the field must also be accounted for.
[56-58]

The second order correlation function of the light is defined by

$$g^2(\tau) = \frac{\langle E^{(-)}(t)E^{(-)}(t+\tau)E^{(+)}(t+\tau)E^{(+)}(\tau)\rangle}{\langle E^{(-)}(t)E^{(+)}(t)\rangle^2},$$

where $E^{(+)}(t)$ and $E^{(-)}(t)$ are the positive and negative frequency
components of the electromagnetic field. The second order correlation function was introduced by Glauber[59] in his formulation of
optical coherence theory. In essence $g^2(\tau)$ is a measure for the
probability that in a light beam after the detection of a photon at
time t a second one is measured at time t+τ. The result of a photon
correlation experiment (Hanbury-Brown and Twiss experiment[60,61])
for a chaotic light source with Gaussian frequency distribution (e.g.
discharge lamp) is a Gaussian with a peak at τ=0 and a width given by
the inverse bandwidth of the light source. That means that there is
a tendency for the photons to arrive in bunches.

If instead of a discharge lamp, a highly stabilized laser is
used for the correlation experiment, one obtains a constant $g^2(\tau)$.
This result holds even when the laser and discharge lamp have the
same band-width. Hence there is some fundamental difference between
a laser and a chaotic light source which may not be apparent in the

spectrum or the first-order correlation function but may, however, be
seen in the second-order correlation function.

The second-order correlation function of the light in resonance
fluorescence has been calculated by Carmichael and Walls and others.
[30,31,33-35] The result at steady state for the saturated atom
($\Omega \gg \Gamma$) is

$$g^2(\tau) = (1 - e^{-3\Gamma\tau/4}\cos \Omega \tau).$$

We see this function exhibits damped oscillations at the Rabi
frequency. Moreover, the extraordinary feature of this correlation
function is that it begins at zero and increases. This is quite
unknown in electromagnetic fields produced by classical sources. The
interpretation of this effect in the phenomenon of resonance fluorescence goes as follows: The first photon detected implies that the
atom has undergone an emission process and so is now in its ground
state. In order to register a correlation, a second photon must be
detected. It is clear that there must be a time lapse for the atom
to regain its excited state. In fact $g^2(\tau)$ is just proportional to
the probability of observing the atom in the upper state when it was
initially prepared in the ground state. We note that as the probability for the atom being in the upper state increases $g^2(\tau)$ may
exceed 1 and we see photon bunching and photon antibunching exhibited
in the same phenomenon. In order to observe the photon antibunching
one requires a very small number of atoms in the observation region
(preferably <1). In the presence of a large number of atoms the
heterodyne signal from the beating of light from different atoms will
completely obscure the photon antibunching in the light from a single
atom.

In a dilute atomic beam with a mean number of atoms in the
observation region <1 one still has the problem of atomic number
fluctuations. As pointed out by Jakeman et al.,[56] the correlation
measurement will include the statistics of the atomic number fluctuations. For Poissonian number fluctuations this implies that g^2
(0)=1. The photon antibunching then cannot be directly observed but
is superposed on the atomic number fluctuations.

So far two photon correlation experiments have been performed on
Na atoms in order to observe the photon antibunching. The first one
has been published by Mandel and coworkers, 57,62,63 the second one
has been carried out in our laboratory. This experiment will be
described in the following.

A dye laser stabilized to a reference atomic beam was used in
the experiment. The standard way of stabilizing a laser is to modu-
late the cavity length. This results in a modulation of the output
frequency, being a disadvantage for the experiment as the effective
linewidth of the laser is broadened in this way. Therefore the

resonance frequency of the reference atomic beam is modulated instead. The reference atomic beam with a collimation ratio of 1:500
was exposed to an oscillating magnetic field produced by a pair of
Helmholtz coils. By introducing an offset $\langle E \rangle \neq 0$ the laser can also
be stabilized to frequencies being slightly off resonance.

For the study of the photon correlation of the fluorescence the
frequency stabilized dye laser beam is directed into the highly
collimated atomic sodium beam (collimation ratio 1:1000) at right
angles. Using circularly pumped light the Zeeman sublevels of the
atoms are optically pumped and after about 100 spontaneous emissions
only the $3^2S_{1/2}, F=2, m=2$ level is populated, which then can be coupled
to the $3^2P_{3/2}, F^1=3, m_F=3$ level. Thus the sodium atoms represent a
two-level system to a good approximation.

In order to be able to observe Rabi oscillations in the photon
correlation, it is essential that the atoms in the observation region
are subject to a constant laser intensity. The spatial distribution
of the laser beam, however, is described by a Gaussian profile.
Therefore, the observation has to be limited to the maximum of the
Gaussian profile, only allowing for laser intensity changes of <10%.
In addition the laser beam is expanded by taking advantage of the
divergence of the dye laser, so that the Gaussian beam profile is
enlarged. Using a collecting lens the fluorescing part of the atomic
beam is magnified by 5:1 and imaged onto an aperture of 1 mm diameter, which transmits only the center part of the fluorescence coming
from the maximum of the Gaussian profile. The atoms have an average
thermal velocity of 4.10^4 cm/sec. As a result the transit time of
the atoms through the observation region is about 500 nsec. The
wings of the Gaussian laser profile, the fluorescence of which is not
observed, already serve to optically pump the atoms. By the time the
atoms enter the observation region, they can be considered as two-
level systems.

In the experiment, the time interval between subsequent fluorescence photons had to be measured. A time to amplitude converter
was used to obtain the nsec time resolution. In order to avoid false
signals produced by photomultiplier ringing, two separate photomultipliers were used for detection, the fluorescent light being
divided by a beam splitter. The photomultiplier pulses were shaped
by a discriminator and then used to start and stop the time to amplitude converter.

The output of the time to amplitude converter was processed in a
multichannel analyzer working in the pulse height analysis mode. The
channel number in which a pulse is stored corresponds to the time
interval τ between start and stop pulses. After averaging, the data
in the multichannel analyzer represent the probability that two
subsequent photomultiplier pulses are separated by the time interval
τ, as a function of τ.

Figure 8 shows three experimental histograms for three different laser intensities with the laser frequency in resonance with the atomic transition frequency. The solid curves result from least squares fit to the theoretical function

$$g^2(\tau) = 1 + \frac{1}{\overline{N}(1+\delta/\overline{N})^2}[1-e^{-3\Gamma\tau/4}(\cos\Omega\tau + \frac{3\Gamma}{4\Omega}\sin\Omega\tau)]$$

to the experimental data. Here $\Omega = \sqrt{\omega^2(\Gamma/4)^2}$, Ω is the Rabi frequency, Γ the natural decay rate, N the average number of atoms in the observation volume and N/δ is the signal to noise ratio. $g^2(\tau)$ is taken from Carmichael et al.,[58] (The number of observed coherence areas is approximately 10^3. The formula is only valid if a large number of coherence areas are observed).

The measurement for 10 mW (Figure 8) shows the Rabi oscillations to level out for longer times to a value larger than the minimum at $\tau=0$, which clearly exhibits the antibunching at $\tau=0$. This is in contrast to the experiments performed by Kimble, Dagenais and Mandel,[57] where the number of coincidences gets smaller rather fast for large delay values due to the short transit time ($\approx$100 nsec) of the atoms through the observation volume. In the present experiment such a decrease has not been observed since the transit time was 500 nsec.

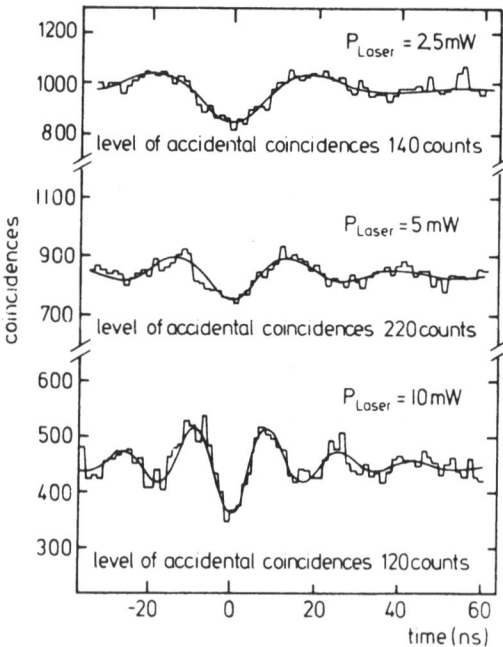

Fig. 8.   Photon correlation of resonance fluorescence of sodium. The smooth curve was obtained by a fit taking into account the number fluctuations of the atomic beam.

By introducing an offset to the modulated magnetic field applied to the reference atomic beam, the laser frequency could be stabilized off resonance. Figure 9 shows experimental curves for the same laser intensity and for the laser frequency on and off resonance. The detuning was $\Delta/2\pi$ = 17 MHz. The change of the Rabi frequency due to the detuning can be seen in both figures. The Rabi frequency for nonresonant excitation is given by

$$\Omega' = (\Omega^2+\Delta^2)^{1/2}$$

where Ω is the Rabi frequency for resonant excitation and Δ the detuning.

According to equation (5) it is obvious that the Rabi oscillations in $g^2(\tau)$ are washed out with increasing average number of N of atoms. In the experiment N has been changed systematically. The influence on $g^2(\tau)$ is shown in Figure 10.

The experiment described here shows clearly the evidence of photon antibunching and thus verifies the predictions of the quantum theory of light.

There is still an interest in the photon correlation at low laser intensities[57] where $\Omega/\Gamma < 1$. In this limit the laser bandwidth changes the photon correlation in a different way compared to the case $\Omega/\Gamma < 1$. The result for low laser intensity gives a generalization of the Heitler-Weisskopf effect applied to photon correlations. The signal is described by

$$g^2(\tau) = 1 + e^{-\Gamma\tau}\frac{1+2\delta/\Gamma}{1-2\delta/\Gamma} - 2\frac{e^{-\Gamma\tau/2-\delta\tau}}{1-2\delta/\Gamma}$$

where δ is the diffusion coefficient of the phase of the laser[57] i.e. the laser linewidth.

For the limit $\delta=0$ (monochromatic source) it follows:

$$g^2(\tau) = (1 - e^{-\Gamma\tau/2})^2.$$

This is reasonable agreement with the measurement shown in Figure 11 when in addition the finite transit time of the atoms through the observation region is considered.

Study of Collisions of Single Atoms

As it was shown in the previous chapter and discussed in the lectures by Hurst, it is possible to detect single atoms. The fluorescence experiments described in the previous chapter show that a

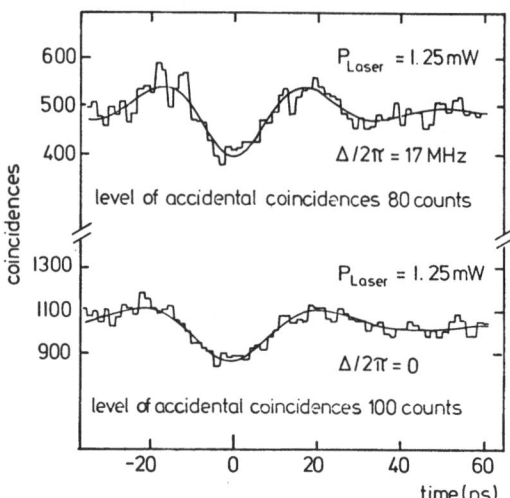

Fig. 9.   Photon correlation of the resonance fluorescence of sodium
          with and without detuning.

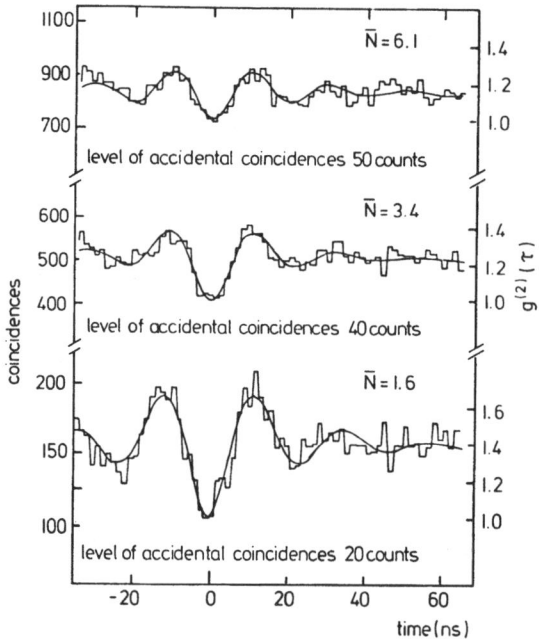

Fig. 10.   Influence of the average atomic number density N on the
           intensity correlation. The oscillations are washed out
           with increasing N.

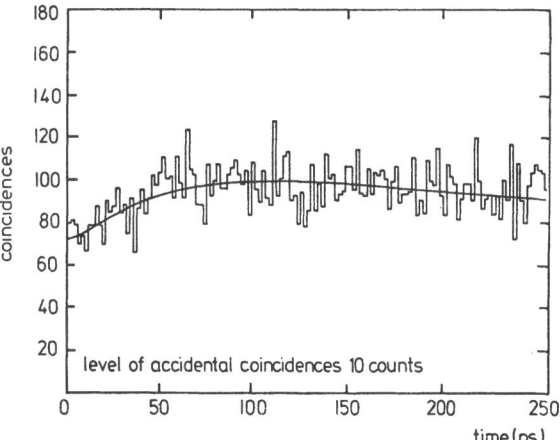

Fig. 11. Photon correlation for low laser intensities $\Omega/\Gamma < 1$. The
data are corrected for finite transit time effects.

detailed understanding of the interaction of an atom with a resonant
laser field can be obtained. In the following an experiment will be
discussed allowing to observe collisions of single atoms.

The setup consists of a proportional counter arrangement as
first described by Hurst et al., (for a review of these techniques
see Ref.[65]). In contrary to their experiments, we do not ionize
the atoms to be detected directly by the laser radiation. First a
Rydberg level (highly excited state of the atom) is populated via a
two step excitation. The free electron necessary for the detection
in the proportional counter is generated in this case by collision
between the rare gas atoms in the counter tube and the Rydberg state.

The two-step resonant excitation to a Rydberg state has now the
advantage that the second excitation step can be tuned to match a
transition from an intermediate state which is different from the
level populated in the first excitation step. This is illustrated
for our experiment on Na atoms in Figure 12 and 13. One laser is
tuned to the $3^2S_{1/2} - 3^2P_{3/2}$ transition, and the second to a line
starting from $3^2P_{1/2}$ to a Rydberg state. It is clear that a signal
in the proportional counter is only observed when energy changing
collisions ($3^2P_{3/2} \rightarrow 3^2P_{1/2}$) occur immediately after laser excitation.

Since the setup provides a one atom sensitivity as was demonstrated in various experiments it is possible to follow the collisional behavior of a single atom e.g. to measure the average time
between excitation and collision, the dependence of the collisions on
alignment or orientation; it is even possible to get an estimate for
the duration of a collision. These are interesting new possibilities
for further new experiments.

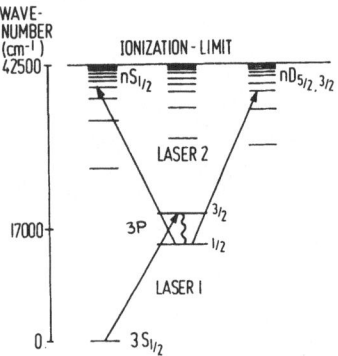

Fig. 12. Excitation steps used to populate the Rydberg states. A
signal pulse is observed in the proportional counter when
energy changing collisions occur in the intermediate state
indicated by the wavy line.

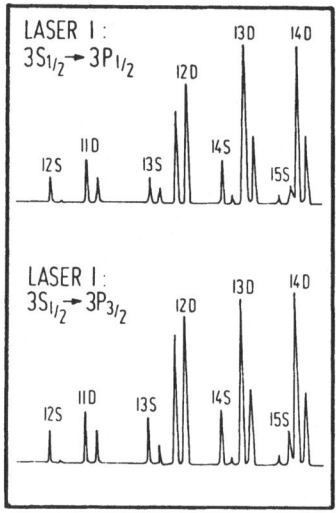

Fig. 13. Signal in the proportional counter when the laser
performing the second excitation step is tuned. The first
excitation step is performed to different fine structure
states in the upper and lower signal curve.

STUDY OF MOLECULE SURFACE INTERACTION BY LASERS

The detailed understanding of surface reactions is of large
technical and scientific interest due to several reasons:

(a) Large scale industrial processes are realized using catalytic
reactions; important examples are ammonia synthesis, oil cracking,
methanol production, catalytic combustion etc.

(b) In material science surface reactions are essential in connection with corrosion, steel hardening, or hydrogen embrittlement.
(c) Reactions between surfaces and electrolyts play a role in elec-
trochemistry e.g. in full cells or in hydrogen production by photo-
lysis.
(d) In biology and medicine surface reactions are of importance for
e.g. bioadhesion or biomembranes.

A large number of methods has been developed to study surfaces.
At the beginning the solid state properties of well defined surfaces
and the adsorbed layers were investigated and a detailed picture of
the atomic positions and electron structure of many adsorption and
coadsorption systems has been obtained. This knowledge about the
static system was the prerequisite for the study of the dynamics of
the interaction and of the surface reactions.

The characterisation of a reaction is obtained studying the
reaction kinetics and the reaction dynamics. The first one is described by the rate constants and the microscopic details of the
reaction whereas the reaction dynamics is described by the reaction
path, the dependence of the reaction cross section on internal and
translational energies of the reactants and on the orientation of the
reactants. The experimental studies of reaction dynamics requires
therefore molecular beam type experiments where no secondary processes as e.g. collisions influence the results. This is not necess-
arily the case for studies of reaction kinetics.

Especially in connection with studies of reaction dynamics the
laser is a useful diagnostic tool. In the following a few experiments of this type will be discussed.

As long as only atoms are involved in the collision process the
measurement of the angular distribution and the momentum change
provides sufficient insight into the scattering dynamics. However,
when molecules are scattered, additional information on the change of
the internal energy is necessary. Recently the laser-induced-fluor-
escence method has been successfully used in several experiments to
determine the influence of the surface interaction on the molecular
rotational distribution.[66-73] It has been shown that for the case
of a carbon-covered Pt(111) surface the rotational degree of freedom
of the scattered NO molecules is only partly accommodated to the
surface temperature.[66-68] A similar effect has also been observed
for molecules which experience only weak inelastic interaction with a
solid surface, such as CO/LiF (001)[69] and NO.Ag(111)[70,71] as well
as for NO molecules desorbing from a Ru(001) surface.[72] In the
following, our own investigations of the angular and rotational
distributions of NO molecules scattered from different surfaces will
be described in more detail.

In the experiments the laser-induced fluorescence of the scattered molecules is measured in the $^2\Sigma \leftarrow {}^2\pi$ (0-0) transition of NO before and after the scattering process. The details of the experimental setup are described in Refs.[59,61,67] The UHV scattering chamber (1-0^{-10}Torr) contained a rotatable quadrupole mass filter for determining the angular distribution of the scattered molecules. A supersonic NO beam with a particle flux of about 2 x 10^{14} molecules cm^{-2}sec^{-1} was scattered from the surface of a Moore-type pyrographite crystal, from a carbon-covered Pt surface, or from a Pt(111) surface. The graphite crystal consisted of microcrystals most of whose c-axes pointed in the direction perpendicular to the surface, and whose a-axes were randomly oriented. The surfaces could be cooled to 130°K with liquid nitrogen and heated via the tantalum support leads. In the following experiments with the graphite crystal will be described in more detail as an example (The experimental setup is shown in Figure 14).

The properties of the graphite surface were probed by scattering a supersonic He beam from the crystal.[75] The full width at half maximum of the specular lobe was twice the width of the incident beam. The peak intensity of the specularly scattered He atoms was only about 9 per cent of the incident beam intensity, which is a small value compared with measurements on graphite single crystals. [76] According to these experiments a relatively high density of surface defects had to be assumed, which were due mainly to grain boundaries and twin crystals with slightly misoriented surface normals.

For temperatures below 700°K chemical reactions between the surface atoms and impurity molecules in the beam (1%, with main constituents N_2O, NO_2, and N_2) could be excluded. An estimate can be given for the NO coverage of the graphite surface during the scattering experiment. From measurements with a chopped NO beam at a surface temperature of 140°K an upper limit for the residence time of 4 x 10^{-5}sec can be deduced. With the given particle flux and a sticking probability of unity, the upper limit of the NO coverage was estimated to be about 10^{-6} monolayers.

The frequency-doubled radiation of an excimer-pumped dye laser was used to probe the NO molecules (5nsec pulses at a rate of 5 Hz, pulse energy 10μJ in the desired wavelength range around 226 nm). Rotational state distributions in the electronic ground state could be derived from the measured fluorescence spectra as described previously.[66]

The electronic ground state of NO is split into two states owing to spin-orbit interaction, with $^2\pi_{1/2}$ state. For the molecules of the incident beam, the measured rotational populations show Boltzmann distributions in both fine-structure states with temperatures corresponding to 40°K ($^2\pi_{3/2}$) and about 70°K ($^2\pi_{1/2}$). The second tem-

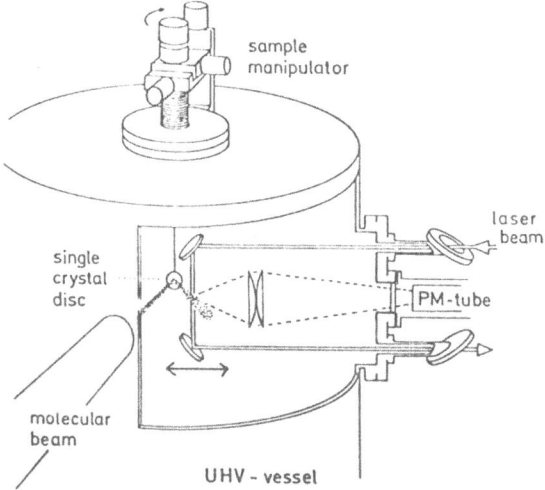

Fig. 14. Experimental arrangement (schematic). The laser beam
          enters and exits the vacuum chamber through Brewster angle
          quartz windows as shown. Inside the chamber it is de-
          flected by aluminum coated mirrors which are displaceable
          left and right as shown, so that the incoming NO beam or
          the scattered molecules exiting the surface could be
          excited and observed.

perature also expresses the ratio of the populations of the two
electronic ground states. This indicates that these states are not
completely equilibrated with each other by the collision processes
during expansion in the nozzle kept at room temperature. Assuming
isenthalpic expansion, one can estimate the translational energy of
the incident molecules from the measured rotational distribution to
be about 700 cm^{-1} or 0.08 eV.

   The rotational and angular distribtuions of the scattered NO
molecules were investigated for surface temperatures between 130°K
and 780°K. Examples of the measured angular distributions are shown
in Figure 15 for incidence angles of 30° and 60° and for different
surface temperatures. The experimental points correspond directly to
the mass spectrometer signals. In order to obtain the angular flux
distribution of the scattered molecules it is necessary to correct
for the velocity of the molecules, which may be different for the
isotropically and specularly scattered molecules. In the present
experiment the velocities of the scattered molecules were not
analyzed, therefore the corresponding corrections could not be
considered.

   In the figure one can distinguish broad scattering lobes in the
direction close to specular reflection, and underlying cosinelike

distributions caused by diffusive scattering. The specular scattering lobe is interpreted as due to weakly inelastic scattering
processes. The half-width of the lobe is about 40° and is independent of the scattering angle. As the surface temperature is raised,
the direction of the lobular maximum approaches an angle to the
surface normal, corresponding qualitatively to the predictions of the
hard cube model.[77,78]

    As shown in Figure 15 as the surface temperature increases, the
fraction of molecules scattered in a cosine distribution decreases
while the specularly scattered fraction grows. A considerable fraction of the observed diffusively scattered particles, however, is
thought to be due to the surface roughness of the crystal used. This
is shown by the remaining isotropic part obtained at the highest
temperatures investigated.

    The rotational distributions were measured for both diffusively
scattered and specularly reflected molecules. In both cases the
rotational temperatures of both electronic ground states were determined. The result is a Boltzmann distribution between the rotational
levels with the same temperature for both fine structure states.

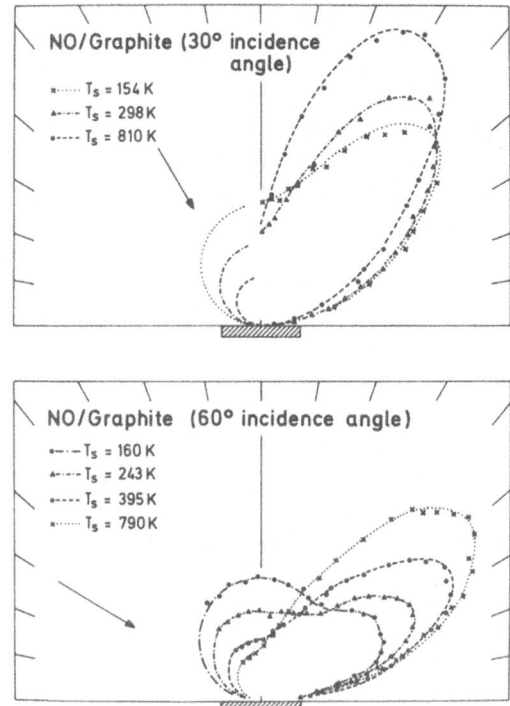

Fig. 15   Angular distributions of NO molecules scattered from a
           graphite surface at different temperatures with incidence
           angles of 30° (upper graph) and 60° (lower graph).

This means that the rotational levels and the two fine structure
states are in thermal equilibrium after the scattering process. This
was the case for all surface temperatures between 130°K and 800°K,
and for both the diffusively scattered and the specularly scattered
molecules.

The dependence of the rotational temperature on the surface
temperature is shown in Figure 16. Each point and the corresponding
error bar is obtained by a least squares fit to the measured population distribution. The solid line corresponds to complete accommodation of the rotational degree of freedom to the surface temperature. The experimental points follow this line up to a surface
temperature of 170°K, they deviate at higher temperatures and from
350°K upwards the rotational temperature converges to a value of
about 250°K.

Most of the rotational distributions were measured for angles of
incidence of 30°. In this case, for the given experimental geometry,
predominantly the specularly scattered molecules were probed by the
laser beam (circles). The same rotational temperatures were obtained
at a second incidence angle of 60° where mainly diffusively scattered
molecules leaving the surface perpendicularly were investigated
(squares). The different experimental geometries, however, imply
different incident energies with respect to the surface normal, which
is important for the interpretation of the experimental data. As is
obvious from the data in Figure 16 the same temperature dependence is
found for both scattering geometries.

Surface scattering experiments are usually discussed in terms of
three basic scattering processes: elastic scattering, inelastic
scattering and trapping/desorption, as characterized by typical
angular distributions of the scattered particles.[79,80] The observation of specularly as well as diffusively scattered NO molecules
may thus be interpreted as a superposition of inelastic scattering
and trapping/desorption processes.

As the surface temperature is raised the fraction of specularly
scattered molecules increases, indicating a predominance of inelastic
scattering at higher temperatures. A possible explanation for this
behavior is the temperature dependence of the residence time τ of the
NO molecules in the attractive potential well of the graphite
surface:

$$\tau = \nu^{-1} \exp(E_d/kT_s),$$

where T_s is the surface temperature and E_d is the desorption-energy of
the NO molecules (estimated to be 0.12 eV for low coverage.[81] The
frequency factor ν is assumed to be 10^{13}sec^{-1}. This equation yields
residence times between $\sim 10^{-9}$ sec for surface temperatures of 150°K,
and ~ 1 psec for $T_s = 700$°K. As the transit time of the molecules

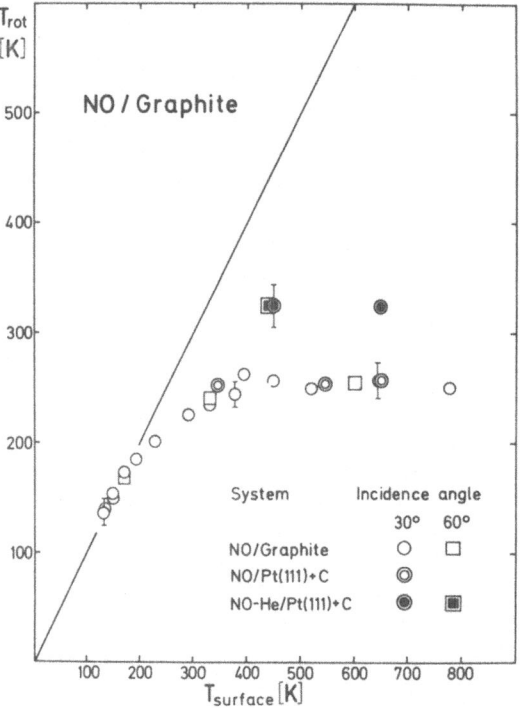

Fig. 16. Plot of the measured rotational temperature of the
scattered NO molecules versus the surface temperature for
different surfaces and incidence angles. The straight line
symbolizes full rotational accommodation.

through the potential well of the surface is estimated to be about 1
psec, the molecules are assumed to undergo many hundred surface
collisions in the case of lower surface temperatures, leading to a
higher fraction of diffusively scattered particles. For higher
temperatures the number of surface collisions decreases, resulting in
a growing specularly scattered fraction.

The essential result we find in the present experiment is that
within the range of rotational energies investigated the rotational
distributions of the scattered molecules correspond to Boltzmann
distributions of the same temperature for both electronic ground
states, and that the ratio of the overall populations of both fine-
structure levels exhibits the same temperature. This only agrees
with the temperature of the graphite surface below 170°K. At higher
surface temperatures there is an increasing discrepancy (Figure 16).

A similar deviation has been obtained in a desorption experiment
where a rotational temperature of 235°K Ru(001) surface, in spite of

a very long residence time.[72]   Assuming that there is no reactive
interaction of the NO/Ru system, the result of the experiment shows
that the final rotational distribution is predominantly determined by
the exit channel of the interaction; the decisive processes leading
to this final rotational distribution seem to take place when the
molecules are receding from the surface.   For the adsorbed molecules,
rotation is hindered by the binding to the surface; thus, rotational
energy is taken up when the molecules desorb from the surface.
Experimental results for NO scattered at Pt(111) surface support this
interpretation[75]:   in spite of residence times of the order of ms
and the observation of primarily diffusive scattering indicating a
trapping/desorption process, rotational temperatures much lower than
the surface temperatures are found for $T_s > 400°K$ (Figure 17).

In the experiment described in this paper we do not have a pure
adsorption/desorption process, since a rather large fraction of the
molecules is specularly scattered.   The result presented in Figure 16
must therefore be due primarily to other processes.

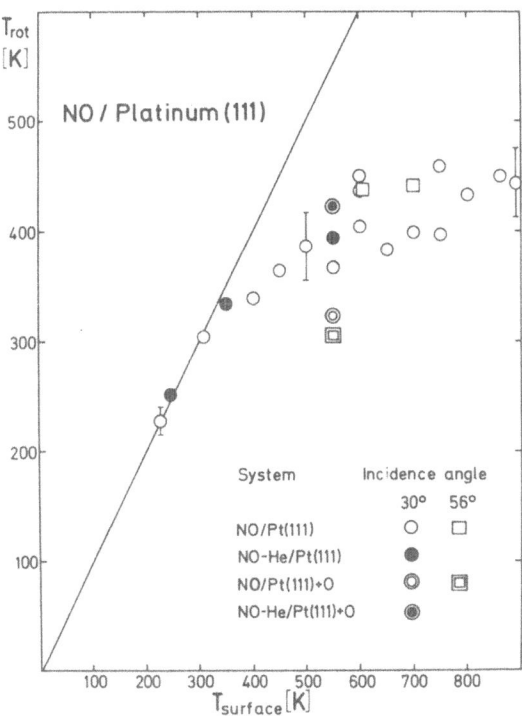

Fig. 17.   Plot of the measured rotational temperature of the
           scattered NO molecules versus the surface temperature for
           different surfaces and incidence angles.   The straight line
           symbolizes full rotational accommodation.

Even at very low surface temperatures the angular distribution shows a specular contribution (Figure 15) indicating molecular residence times comparable to those at high surface temperatures. That means that the measured rotational accommodation at lower temperatures (Figure 16) cannot strongly be influenced by the residence time of the molecules on the surface. It is most likely that in the region of surface temperatures higher than 400°K, the rotational temperature of the scattered molecules is influenced by the kinetic energy of the incoming molecules and that a redistribution of the energy into the different degrees of freedom of the molecules takes place during the interaction with the surface. This is supported by experiments in which a seeded NO beam is scattered from a graphite surface. In this case, for surface temperatures larger than 400°K, a higher rotational temperature has been observed (Figure 16); a corresponding result has not been obtained in the scattering of NO from Pt(111) as is shown in Figure 17. It is quite clear that the description of the processes as given here must be rough and incomplete and that a perfect understanding of the dynamics of the molecule-surface interaction can only be obtained by trajectory calculations.

REFERENCES

1. D. Hinkley, ed., in: "Laser Monitoring of the Atmosphere", Topics in Applied Physics Vol.14, Springer Verlag Berlin, Heidelberg, New York (1976).

2. D. K. Killinger and A. Mooradian, eds, in: "Workshop on Optical and Laser Remote Sensing", Springer Series in Optical Sciences, Springer Verlag Berlin, Heidelberg, New York (1982).

3. K. W. Rothe, H. Walther, in: "Tunable Lasers and Applications", A. Mooradian, T. Jaeger, P. Stokreth, eds Springer Series in Optical Sciences Vol.3, Springer Verlag Berlin, Heidelberg, new York (1976).

4. R. L. Byer and L. A. Shepp, Optics Lett. 4:75 (1979).

5. U. Platt, D. Perner, and H. W. Patz, Journ.Geophys.Rev. 84:6329 (1979).

6. D. Perner and U. Platt, Geophys.Rev.Lett. 6:917 (1979).

7. U. Platt, D. Perner, A. M. Winer, G. W. Harris and J. N. Pitts. Jr., Geophys.Rev.Lett. 88:112 (1983).

8. R. T. Menzies and M. S. Shumate, Science 184:570 (1974).

9. R. T. Menzies in: "Laser Spectroscopy III", J. L. Hall and J. L. Carlsten, eds., Springer Series in Optical Sciences, Springer Verlag Berlin, Heidelberg, New York (1977).

10. V. E. Derr, M. J. Post, R. L. Schwieso, R. F. Calfee, and G. T. McNice, A theoretical Analysis of the Information Contend of LIDAR Atmosphere Returns, NOAA Technical Report ERL 296 - WPL 29.

11. S. Svanberg in: "Surveillance of Environmental Pollution and Resources by Electromagnetic Waves - Principles and Appli-

cations", T. Lund, ed., Nato Advanced Study Institute Series, D. Reichel Publishing Company, Dordrecht, Holland (1978).

12. M. C. Sandford and A. J. Gibson, J.Atmosph.Terr.Phys. 32:1423 (1970).
13. A. J. Gibson and L. Thomas, Nature 256:561 (1975).
14. K. W. Rothe, U. Brinkmann, and H. Walther, Appl.Phys. 3:114 (1974).
15. K. W. Rothe, U. Brinkmann, and H. Walther, Appl.Phys. 4:181 (1974).
16. R. L. Byer and M. Garbuny, Appl.Optics 12:1496 (1973).
17. G. Mégie, J. Y. Allain, M. L. Chanin, and J. E. Bamont, Nature 270:329 (1977).
18. M. L. Chanin, J. W. Allain, and J. Pelon, Ninth International Laser-Radar Conference, Munich, July (1979).
19. M. J. Molina and F. S. Rowland, Nature 249:810 (1974).
20. O. Uchino, M. Maeda, and M. Hirono, IEEE, QE-15, 10:1094 (1979).
21. J. C. Petheram Applied Optics 20:3951 (1981).
22. D. K. Killinger, and N. Menyuk, IEEE J. of Quantum Electronics, QE-17:1917 (1981).
23. M. S. Shumate, R. T. Menzies, W. B. Grant, and D. S. Dougal, Applied Optics, 20:545 (1981).
24. W. Baumer, Ph. D. thesis, Fakultät für Physik, Ludwig-Maximilians-Universität Müchen, July (1979).
25. B. R. Mollow, Phys.Rev. 188:1969 (1969).
26. G. Oliver, E. Ressayre, and A. Tallet, Nuovo Cimento Lett. 2:77 (1971).
27. B. R. Mollow, Phys.Rev. 165:145 (1975).
28. S. S. Hassan and R. K. Bullough, J.Phys. B8:L147 (1975).
29. S. Swain, J.Phys. B8:L437 (1975).
30. C. Cohen-Tannoudji, in: "Proc. 2nd Int. Laser Spectroscopy Conf.", Mégève, France, 1975, S. Haroche, J. C. Pebay-Peyroula, T. W. Hänsch, S. E. Harris, eds., Springer Berlin, Heidelberg, New York, p. 324 (1975).
31. H. J. Kimble and L. Mandel, Phys.Rev. A 13:2123 (1976).
32. K. Wodkiewicz and H. H. Eberly, Ann.Phys. (N.Y.) 101:514 (1976).
33. H. J. Carmichael and D. F. Walls, J.Phys. B8:L77 (1975).
34. H. J. Carmichael and D. F. Walls, J.Phys. B9:L43 (1976).
35. H. J. Carmichael and D. F. Walls, J.Phys. B9:1199 (1976).
36. R. J. Ballagh, PhD Thesis Univ. of Colorado, USA, (1982).
37. J. D. Cresser, PhD Thesis Univ. of Queensland, Australia, (1981).
38. F. Schuda, C. R. stroud, Jr. and M. Hercher, J.Phys. B1:L198 (1974).
39. H. Walther, "Proc.2nd Int. Laser Spectroscopy Conf.", Mégève, France, 1975, S. Haroche, J. C. Pebay-Peyroula, T. W. Hänsch, S. E. Harris, eds., Springer Berlin, Heidelberg, New York, p.358 (1975).
40. F. Y. Wu, R. E. Grove, and S. Ezekiel, Phys.Rev.Lett. 35:1426 (1975).

41. W. Hartig, W. Rasmussen, R. Schieder, and H. Walther, Z.Phys. A 278:205 (1976).

42. R. E. Grove, F. Y. Wu, and S. Ezekiel, Phys.Rev. A. 15:227 (1977).

43. H. Carmichael and D. F. Walls, J.Phys. B10:L685 (1977).

44. J. D. Cresser, J. Häger, G. Leuchs, M. Rateike and H. Walther in: "Dissipative Systems in Quantum Optics", R. Bonifacio, ed., Springer Berlin, Heidelberg, New York, p.21 (1982).

45. W. Heitler, "Quantum Theory of Radiation" 3rd Edition, London: Oxford Univ.Press (1964).

46. P. A. Apanasevich, Optics and Spectroscopy 16:387 (1964).

47. P. A. Apanasevich, Optics and Spectroscopy 14:324 (1963).

48. M. C. Newstein, Phys.Rev. 167:89 (1968).

49. M. Lax, Phys.Rev. 157:213 (1967).

50. H. Paul and W. Brunner, Optical Acta, 27:263 (1980).

51. H. D. Simeon and R. Loudon, J.Phys. A 8:539 (1975).

52. K. J. M'Neil and D. F. Walls, J.Phys. A 7:617 (1974).

53. A. Bandilla and H.-H. Ritze, Opt.Comm. 28:126 (1979).

54. H.-H. Ritze and A. Bandilla, Opt.Comm. 28:241 (1979).

55. D. Stoler, Phys.Rev.Lett. 33:1397 (1974).

56. E. Jakeman, E. R. Pike, P. N. Pusey, and J. M. Vaughan, J.Phys. A 10:L257 (1977).

57. H. J. Kimble, M. Dagenais, and L. Mandel, Phys.Rev. A 18:201 (1978).

58. H. J. Carmichael, P. Drummond, P. Meystre, and D. F. Walls, J.Phys. A 11:L121 (1978).

59. R. J. Glauber, Phys.Rev. 130:2529 (1963) and 131:2766 (1963).

60. R. Hanbury-Brown and R. Q Twiss, Nature 177:27 (1956).

61. R. Hanbury-Brown and R. W. Twiss, Proc.Roy.Soc. A 242:300 (1957); A 243:291 (1957).

62. H. J. Kimble, M. Dagenais, and L. Mandel, Phys.Rev.Lett. 39:691 (1977).

63. M. Dagenais, and L. Mandel, Phys.Rev. A 18:2217 (1978).

64. K. Wodkiewicz, Phys.Lett. A 77:315 (1980).

65. G. S. Hurst, M. G. Payne, S. D. Kramer, and D. H. Chen, Phys.Today 33:24 (1980).

66. F. Frenkel, J. Häger, W. Krieger, H. Walther, C. T. Campbell, G. Ertl, H. Kuipers, and J. Segner, Phys.Rev.Lett. 46:152 (1981).

67. G. M. McClelland, G. D. Kubiak, H. G. Rennagel, and R. N. Zare, Phys.Rev.Lett. 46:831 (1981).

68. F. Frenkel, J. Häger, W. Krieger, H. Walther, C. T. Campbell, G. Ertl, H. Kuipers, and J. Segner in: "Laser Spectroscopy V", A. R. W. McKellar, T. Oka, B. P. Stoicheff, eds., Springer, Berlin, Heidelberg, New York, p. 425, (1981).

69. J. W. Hepburn, F. J. Nothrup, G. L. Ogram, J. C. Polanyi, and J. M. Williamson, Chem.Phys.Lett. 85:127 (1982).

70. A. W. Kleyn, A. C. Luntz, and D. J. Auerbach, Phys.Rev.Lett. 47:1169 (1981).

71. A. C. Luntz, A. W. Kleyn, and D. J. Auerbach, J.Chem.Phys. 76:737 (1982).

72. R. R. Cavanagh and D. S. King, Phys.Rev.Lett. 47:1829 (1981).
73. L. D. Talley, W. A. Sanders, D. J. Bogan, and M. C. Lin, Chem.Phys.Lett. 66:500 (1981).
74. F. Frenkel, J. Häger, W. Krieger, H. Walther, G. Ertl, J. Segner, and W. Kielbuber, Chem.Phys.Lett. 90:225 (1982).
75. D. L. Smith and R. P. Merrill, J.Chem.Phys. 52:5861 (1970).
76. G. Boato, P. Cantini, C. Guidi and R. Tatarek, Phys.Rev. B 20: 3957 (1979).
77. R. M. Logan and R. E. Stickney, J.Chem.Phys. 44:195 (1966).
78. W. L. Nichols and J. H. Weare, J.Chem.Phys. 63:379 (1975).
79. W. H. Weinberg and R. P. Merrill, J.Chem.Phys. 56:2881 (1971).
80. W. H. Weinberg and R. P. Merrill, J.Vac.Sci.Technol. 8:718 (1971).
81. C. E. Brown and D. G. Hall, J.Colloid Interface Sci. 42:334 (1973).

CONTRIBUTED PAPERS:

IR AND UV FREE ELECTRON LASER SOURCES

A. Marino

ENEA, Centro Ricerche Energia
Casaccia
Roma, Italy

The characteristics and developments of free electron laser
sources have been discussed in this lecture. After a brief review of
the theoretical aspects of this new laser source, the experimental
proposals and the first obtained results have been reported both for
single pass electron beam devices (IR-FIR) and for storage ring
recirculated electron beam (VUV and XUV). The wide range of tunability and the large attainable power make the FEL a unique source of
coherent electromagnetic radiation in the submillimeter wavelength
range.

Particular attention has been devoted to the electron beam
source and to the wiggler magnet since the laser beam properties
strongly depend on these parameters. Finally, the ENEA Frascati FEL
project has been illustrated.

STUDY OF ASBESTOS BY LASER MICROPROBE

MASS ANALYSIS (LAMMA)

J. De Waele and F. Adams

Universitaire Instelling Antwerpen
Wilryk
Belguim

It is known for many years that exposure to asbestos under certain conditions is correlated with lung cancer and asbestosis[1]. The toxicity of the asbestos fibers is believed to be associated with the surface adsorption power and reactivity of these minerals. Therefore it is desirable to develop methods capable of identifying organic material adsorbed on individual asbestos fibers and to investigate the behavior of the surface after different treatments.

The laser microprobe mass analyser operates with a Nd:YAG laser generating 15 nsec light pulses with a wavelength, after frequency quadrupling, of 265 nm. The laser is focused on the specimen, using an optical microscope. The ions formed by the laser irradiation of a small sample area of approximately $1\mu m^2$ are accelerated into a time-of-flight mass spectrometer (TOF-MS). The mass analysed ions are detected with an electron multiplier and the signals are stored in a fast 8 bit transient recorder. The laser and the TOF-MS are at 180°. The laser energy can be varied with a set of attenuation filters. A more detailed description of the LAMMA-500 instrument (Leybold-Heraeus GmbH, Köln, FRG) is described elsewhere[2,3].

The asbestos minerals studied (chrysotile A and B, crocidolite, amosite and anthophyllite) were supplied by the UICC (Union Internationale Contre le Cancer)[4]. To introduce asbestos into the LAMMA instrument for analysis, a Formvar coated electron microscope grid is brought in contact with the fibers. A number of them stick to the grid coating and can be analysed individually.

In order to study the capabilities of LAMMA to detect organic compounds present at the surface of individual fibers, several pure species were adsorbed onto the different types of asbestos.

Benzo(a)pyrene (BaP) was adsorbed by refluxing 30 mg of crocidolite
in a 10^{-3} M BaP solution in benzene during 15 hours, and by gas-phase
adsorption using an expanded-bed adsorption tub[5]. Benzidine was
adsorbed from a 10^{-3} M benzidine dihydrochloride solution in water at
60° for 19 hours. N, N-dimethylaniline (DMA) was adsorbed from
benzene solution onto the asbestos samples which had previously been
treated with 0.1 N HCl. In all cases the concentration of the organic impurities amounted to ~1 mg g^{-1}.

The application of laser desorption conditions resulted in the
easy detection of benzo(a)pyrene and benzidine, e.g., through their
molecular ion peak (M)$^+$. In addition, in certain conditions organic
contaminants were detected at the asbestos surface, probably anti-
oxidants from polyethylene used for storing the samples. The mass
spectra taken at high laser energy show considerable fragmentation
spectra of both the organic compounds and the asbestos substrate.
However, for the latter contribution the mass spectra are significantly different after treatment compared to those of the raw fibers,
suggesting leaching of magnesium and iron. In accord with literature
data[6], it appeared that anthophyllite adsorbs the organic material
much less readily than the other asbestos types.

For dimethylaniline (DMA) a number of surface catalysed oxidation products (e.g., Methyl Violet) are detectable in the mass
spectra. Figure 1 shows the positive spectra of Methyl Violet and

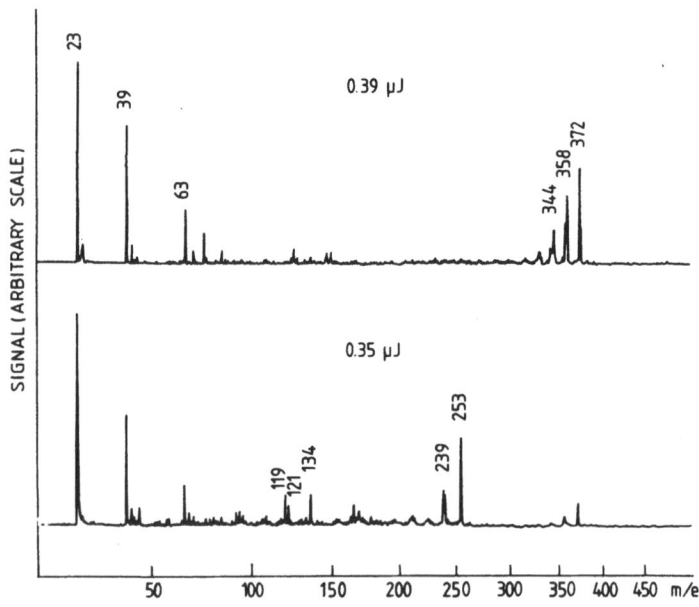

Fig. 1.   Positive laser desorption mass spectra of Methyl Violet
(top) and dimethylaniline (bottom) adsorbed onto
crocidolite.

DMA adsorbed onto crocidolite. The mass peaks at m/e = 372 and 358 of Methyl Violet and those at m/e = 253, 239 and 134 correspond with fragments which can all be identified as specific reaction products. The mass peak at m/e = 121 corresponds with unreacted DMA. LAMMA hence corroborates on the microscopic level earlier findings in which the catalytic oxidation of DMA at the surface of kaolinite, sepiolite and other zeolites was proven[7]. Again anthophyllite is exceptional and does not show any oxidation products.

It appears that LAMMA is able to detect organic components adsorbed onto asbestos surface, at a concentration level of the order of 1000 μg g^{-1}. The technique offers a potential for the characterization of raw and industrially processed asbestos fibers and may hence become of considerable importance to resolve pertinent questions on the toxicity of these compounds. It appears repeatedly in the course of this work that the asbestos standards used, and also several other samples which have been investigated, contain organic surface contaminants, thus reflecting the important surface reactivity. The reaction products formed after adsorption of N,N-dimethyl-aniline stress the catalytic oxidative behavior of several types of asbestos fibers.

Acknowledgements

The authors are affiliated with the Department of Chemistry, University of Antwerp (U.I.A.), B-2610 Wilrijk, Belgium. Support from IWONL (Belgium) and NATO is acknowledged.

REFERENCES

1. P. C. Elmes, R.Soc.Hlth.J., 96:248-252 (1976).
2. E. Denoyer, R. Van Grieken, F. Adams, and D. F. S. Natusch, Anal.Chem., 54:26A-41A (1982).
3. H. Vogt, M. J. Heinen, S. Meier, and R. Wechsung, Fres.Z.Anal. Chemie, 308:195-200 (1981).
4. V. Timbrell, J. C. Gilson, and I. Webster, Int.J.Cancer, 3: 406-408 (1968).
5. A. H. Miguel and D. F. S. Natusch, Anal.Chem., 47:1705-1707 (1975).
6. M. C. Markham and K. Wosczyna, Environm.Sci.Technol., 10:930-931 (1976).
7. E. F. Vansant and S. Yariv, J.Chem.Soc.,Far.Trans.I., 73:1815-1824 (1977).

MATRIX-ISOLATION SPECTROSCOPY OF SMALL METAL CLUSTERS

H. Sontag, B. Eberle and R. Weber

Fakultät für Physik, Universität Konstanz
Konstanz
Fed.Rep.Germany

In the last few years, the interest in the properties of small metal clusters has grown rapidly because of their possible application in the field of catalysis. The spectroscopy of such particles isolated in noble gas matrices[1] offers a considerable reduction in the complexity of their molecular spectra and gives valuable information on the gas phase properties. In addition particles which are produced only a low rates can easily be collected for spectroscopic measurements. The matrix isolation (MI) technique is therefore especially suitable for the spectroscopy of small clusters.

Other advantages of the MI-technique include the following:

- all particles are in a fixed position in the lowest thermal state while in gas phase spectroscopy usually only a very small fraction is in a single vibronic state. The detection sensitivity is therefore greatly enhanced;
- spectra are simplified by the absence of rotational structure;
- fluorescence can be quenched by nonradiative processes. In these spectral regions Raman scattering can then be performed.

The disadvantages of MI-spectroscopy include the loss of information on the rotational structure, the limited resolution and matrix-effects. These matrix effects however are usually fairly low and the spectroscopic constants derived agree within 1% with gas phase values. The resolution is limited by inhomogeneous broadening and therefore depends strongly on the preparation and treatment of the matrix. The linewidths of fluorescence lines is typically in the order of 1-3 cm^{-1}, while the linewidth of Raman lines can be much less than 1 cm^{-1}.

Most publications deal with the observation of dimers, [see for example Ref.1]. Spectra are usually obtained by absorption measurements, laser-induced fluorescence and laser excitation spectra or by resonance Raman scattering. Only few reports of larger clusters are available. Ag_3 and Ni_3 have been observed by resonance Raman scattering[2,3]. For Ni_3 the structure in the matrix has been derived from the isotope effect. Bi_4 has been analyzed by laser-induced fluorescence[4]. In this case the complete vibrational spectrum and excited electronic states were measured. Sb_4 has been observed by non-resonant Raman scattering[1]. In this case all vibrational modes were observed too.

The spectroscopy of similar and larger clusters will strongly depend on efficient preparation procedures of these particles.

Acknowledgement

Financial support of the Deutsche Forschungsgemeinschaft is gratefully acknowledged.

REFERENCES

1. H. Sontag and R. Weber, Chem.Phys., 70:23 (1982).
2. W. Schulze, H. U. Becker, R. Minkwitz, and K. Manzel, Chem.Phys. Letters, 55:59 (1978).
3. M. Moskovits and D. P. Di Lella, J.Chem.Phys., 72:2267 (1980).
4. V. E. Bondybey and J. H. English, J.Chem.Phys., 73:42 (1980).

THE USE OF RESONANCE RAMAN SPECTROSCOPY AS AN ANALYTICAL TOOL IN THE QUALITATIVE AND QUANTITATIVE DETERMINATION OF SOME FOOD DYES

R. F. Stobbaerts and M. A. Herman

Laboratory for Inorganic Chemistry
Rijksuniversitair Centrum Antwerpen
Antwerp, Belguim

The use of resonance Raman spectroscopy was studied in the determination of all yellow, orange and red food dyes registered by the European Economical Community. This technique offers many advantages in comparison with the commonly used methods, mostly based on chromatography or visible absorption spectroscopy.

The natural dyes studied do not exhibit a resonance Raman spectrum, but most of the artificial dyes give spectra that can be used for analytical determinations.

It was found that for all but one of the dyes considered, the 488 nm Ar^+-laser line yields optimal spectra that differ to the extent that they can be used as a fingerprint of the dye present.

Taking the identification limit as being the concentration for which at least five bands appear in the spectrum, ranging from 800 cm^{-1} to 1900 cm^{-1}, with a signal-to-noise ratio of at least 3, the values found vary between 100 and 150 ppb. Detection was still possible for solutions with a concentration as low as 50-60 ppb.

Using the aqueous solutions of the colored outer layer of a yellow and a green bubble gum in parallel experiments with UV/VIS spectroscopy, it turned out that only the resonance Raman spectra were able to allow exclusive identification of the coloring dye. In the case of the green bubble gum, it was possible, by using the excitation frequency of 488 nm, to record the spectrum of a yellow-compound being present, thereby proving that the green color had been obtained by mixing a yellow and a blue dye, the latter not responding to the resonance criteria so that its spectrum almost disappeared in the background noise. In both cases the yellow compound could easily

be identified as tartrazin, registered under CI 19140 in the Color
Index classification.

Using water as an internal standard, giving rise to a broad band
at about 1640 cm^{-1}, a calibration plot based on relative intensities
was worked out for CI 19140, yielding an accuracy of approximately
5-6% when applied to the data obtained from the spectra of the solutions of the colored layers from the bubble gum samples. In order
to improve the accuracy the spectra then were considered to be composed of the spectrum of the dye, the spectrum of the solvent and a
variable base-line. A computer-program, based on multiple linear
regression analysis, was written to calculate the correlation coefficients, besides the value of the statistical F-test, and the chi-
square value to evaluate the goodness of fit. Intensive work is
still being performed on this matter, but the preliminary results are
very promising.

The use of water as an internal standard may give rise to some
difficulties in cases where the spectrum of the dye investigated
shows some bands that interfere with the broad band, due to the
presence of water, at about 1640 cm^{-1}. To eliminate this, another
internal standard was sought. It was found that recording the
spectra of the dyes in 0.25 M BO_3^{3-} gives a supplementary sharp,
intense peak at 879 cm^{-1}, thus in a region that is almost band-free
for all dyes investigated. Further experiments showed that none of
the dyes react with the borate anion, even after long exposure to
high-power laser irradiation[1].

The proposed technique thus offers many advantages by solving
various problems that occur by using the more commonly used methods
in this field[2,3].

Acknowledgements

The authors wish to thank L. Van Haverbeke for his many interesting discussions on the subject. One of us (R. F. Stobbaerts)
wants to thank the NATO-ASI for financial support to attend the
course.

REFERENCES

1. R. F. Stobbaerts, L. Van Haverbeke, and M. A. Herman, J.Food
    Sci., 48:521 (1983).
2. C. W. Brown and P. F. Lynch, J.Food Sci., 41:1231 (1976).
3. S. Higuchi, J. Tanaka, and S. Tanaka, J.of the Spectr.Soc.
    of Japan, 27:353 (1978).

APPLICATION OF THE R.I.S. TECHNIQUE TO THE

STUDY OF IMPURITIES IN SOLIDS

E. Arimondo, S. Martellucci, E. Santamato,
A. Sasso, and S. Solimeno

University of Naples
Italy

Resonance ionization spectroscopy is based on directly detecting the ion-electron pair generated in single- or several-photon laser absorption processes. By using lasers tuned on atomic transitions and proportional counters as single-electron detectors, the possibility has been demonstrated of detecting a single atom, present in the laser interaction volume, both in the case of experiments in the vacuum and in the case when a buffer gas is present. This technique enables minute traces of alkaline, alkaline-earth, certain rare-earth, and some inert gas atoms to be detected. It may be used for high precision microanalysis, which finds significant applications in the study of surface phenomena.

As regards the investigation of properties of surfaces, RIS can yield information not obtainable by other methods:

a) compared with secondary ions mass spectrometry (SIMS), where secondary ions engendered in surface sputtering are analysed with a mass spectrometer, RIS enables the study of neutral species, emitted at the same time of ions, thus yielding supplementary informations;
b) RIS enables the temporal evolution of neutral species deadsorbed from a surface to be followed, thus directly monitoring the atoms present on a surface by means of their time variations;
c) RIS may also be applied to simple molecules such as NO and CO for which surface-physical data are scarce.

The aim of this lecture has been to review the experimental approach used for implementing such a technique and to compare it with some more conventional approach based on mass spectrometry and atomic and molecular beam scattering.